Multiphase Polymer-Based Materials

An Atlas of Phase Morphology at the Nano and Micro Scale

Multiphase Polymer-Based Materials

An Atlas of Phase Morphology at the Nano and Micro Scale

Charef Harrats

CRC Press
Taylor & Francis Group
Boca Raton London New York

CRC Press is an imprint of the
Taylor & Francis Group, an **informa** business

CRC Press
Taylor & Francis Group
6000 Broken Sound Parkway NW, Suite 300
Boca Raton, FL 33487-2742

First issued in paperback 2019

© 2009 by Taylor & Francis Group, LLC
CRC Press is an imprint of Taylor & Francis Group, an Informa business

No claim to original U.S. Government works

ISBN-13: 978-1-4200-6217-5 (hbk)
ISBN-13: 978-0-367-38587-3 (pbk)

Library of Congress Cataloging-in-Publication Data

Harrats, Charef.
    Multiphase polymer- based materials : an atlas of phase morphology at the nano and micro scale / Charef Harrats.
        p. cm.
    Includes bibliographical references and index.
    ISBN 978-1-4200-6217-5 (alk. paper)
    1. Polymers. 2. Nanostructured materials. 3. Phase transformations (Statistical physics) I. Title.

QC173.4.P65H367 2008
668.9--dc22   2008044155

Visit the Taylor & Francis Web site at
http://www.taylorandfrancis.com

and the CRC Press Web site at
http://www.crcpress.com

Dedication

I dedicate this book to my wife, Ralisa, and my daughters, Meriem, Nejda, Hinde, and Watitou.
I also dedicate the work to my two sons, Kais (age 16) and Mohamed-Wassim (age 4).
Although it was very funny to see Mohamed-Wassim play
in my home office, he was also a disturbance.

To Professor Gabriel Groeninckx, thank you very much for the scientific, education,
and social atmospheres you have created for us in your Laboratory for
Macromolecular Structural Chemistry at the Katholieke Universiteit Leuven
during our fruitful collaboration in the period from 1997 to 2005.

You were not only the scientist we know, but a very sensitive and warm person to your
collaborators. The door to your office was always open for us.
You shared everything you could with your students, postdocs, and associates.
You encouraged and financed even first year Ph.D. students to participate in seminars, colloquium,
and congresses worldwide. Very rare are those who do so. We won't forget that.

My present modest words would not be enough to fully express
my acknowledgments and recognition to you.

Contents

Preface ix

Acknowledgments xi

About the Authors xiii

1   Basics in Phase Morphologies of Multicomponent Polymer-Based Materials 1
1.1   Phase Morphology Development in Polymer Blends 1
    1.1.1   Deformation, Breakup, and Coalescence Phenomena during
        Melt-Blending 1
        1.1.1.1   Basic Principles 1
    1.1.2   Effect of Component Characteristics on Phase Morphology 3
        1.1.2.1   Viscosity Ratio 3
        1.1.2.2   Blend Composition 3
        1.1.2.3   Effect of Elasticity 5
        1.1.2.4   Role of the Mixing Equipment and Compounding Conditions 6
    1.1.3   Main Types of Phase Morphologies 7
        1.1.3.1   Droplet-in-Matrix Phase Morphology (Dispersed Morphology) 7
        1.1.3.2   Composite Droplet Phase Morphology 10
        1.1.3.3   Cocontinuous Phase Morphology 10
References 13

**2   Phase Morphology Investigation: Microscopic Tools, Tips, and Selected
    Scanning Electron Photomicrographs** 17
2.1   Introduction 17
2.2   Sample Preparation for Scanning Electron Micrography 18
References 31

3   Selected Microscopic Illustrations of Phase Morphology in Nanocomposites 33
3.1   Introduction 33
3.2   Some Specific Aspects Related to the Preparation of Polymer Nanocomposites 33
3.3   Mobility of Polymer Molecules in Confined Spaces of Layered Silicates 35
References 120

**4   Illustrations of Phase Morphology Other Than Cocontinuous
    in Polymer Blends Including Thermoplastics/Thermoplastics and
    Thermosets/Thermoplastics** 123

**5   Scanning Electron Microscopy Illustrations of Two-Phase Cocontinuous
    Morphologies in Binary Polymer Blends** 221

Index 327

Preface

Phase morphology in multicomponent polymer-based systems—including thermoplastics/thermoplastics, thermoplastics/thermosets and thermoplastics/rubbers (called thermoplastic vulcanizates), and polymer matrices where inorganic or organic fillers are dispersed via various processing strategies—constitutes the core physical characteristic to carefully control when designing and studying plastic materials. A huge volume of scientific literature and industrial patents is continuously devoted to research where the interrelation of phase morphology with physical and chemical properties of the materials is considered in controlling the performance of polymer-based materials.

In 2006, I coedited with T. S. Thomas and G. Groeninckx a book on the micro- and nanostructured multiphase polymer blends systems titled *Micro- and Nanostructured Multiphase Polymer Blend Systems: Phase Morphology and Interfaces*. The focus of that book was on the research achieved in multiphase polymer blends, with an emphasis on both theoretical and experimental aspects of their formation, their properties, and their processing in relation to the parameters used in their control.

This book, which can be considered as a complementary work to the 2006 book, utilizes a more practical and straightforward approach in dealing with the microscopic observation of phase morphology in multicomponent polymer blends and nanocomposites. It deals exclusively with the study of phase morphology of materials made of one, two, or more polymers, polymers and nonpolymers, copolymers, and combinations thereof. It targets students, technicians, and phase morphology analysts as it explains the various methods by which phase morphology can be directly observed. It also targets more experienced researchers and professionals as it gives a huge variety of good-quality photographs and illustrations provided by authors accumulating a large expertise in the field.

As it is not the objective of this book to include a detailed discussion of the images, a succinct and brief description is given in addition to the legend for each photomicrograph. In some cases, extended information is provided to help the reader clearly understand the content and features of the images. This is the case where many photomicrographs are grouped in one figure and compared. However, a deeper understanding requires the reading of the source of the photograph. For that, each image is associated with the full reference of the article where extended discussions are given.

Chapter 1 provides a brief introduction on the basics and fundamentals in phase morphology of polymer blends. This topic is frequently reported in a very extended way in open literature, usually in the reviews and also in the books and articles dealing with phase morphology in polymer blends. In Chapter 2, the microscopic tools, tips, and some selected scanning electron photomicrographs are provided as illustrations to the practical methods described for the investigation of morphology. The images presented in Chapter 3 contain morphological features of nanocomposites constituting a variety of polymer matrices where silicates clay was dispersed in an exfoliated or intercalated state. Chapter 4 relates photomicrographs of thermoplastics/thermoplastics, thermoplastics/thermosets, and some selected thermoplastic vulcanizates in which the phase morphology is not particularly cocontinuous. Chapter 5 is fully devoted to microscopic photomicrographs of blend systems where the phase morphology is of the cocontinuous type.

I would like to emphasize that the present book could be of encyclopedia size, as more than one million microscopic pictures on phase morphology of polymer blends are available in literature and are excellent candidates to integrate herein. However, difficult decisions were made in the selection of images. On writing this, I would like to apologize to the many scientists and authors who have published excellent work in this area and that either I passed over by ignorance or I could not really include as I had made a selection of work similar or closer in content to their own.

Charef Harrats
Liege, Belgium

Acknowledgments

I would like to thank Professor Gabriel Groeninckx, one of my two coeditors of our 2006 book on phase morphology in micro- and nanostructured multiphase polymer blend systems, for his encouragement and advice as I prepared this book.

I also thank Dr. T. S. Omonov, the author of the Ph.D. thesis on phase morphology of polymer blends carried out at MSC Laboratory at Katholieke University Leuven. Professor Groeninckx and I supervised his thesis with pleasure, aware of the author's skillful research. A large number of excellent-quality pictures from this thesis are used in this book.

I should not forget all the scientists who provided me with their excellent-quality microscopy images on phase morphology.

About the Author

Charef Harrats, Ph.D., obtained his engineering degree in plastics in 1986 from the Algerian Institute of Petroleum (IAP) in Boumerdes (Algeria). The plastic and rubber department at IAP was a cofoundation of the Algerian Ministry of Energy and the Michigan Institute of Technology in New Jersey (USA). He worked as a process engineer at the National Company for Rubber and Plastics (ENPC-Algeria) and then moved to the University of Liège in Belgium, where he worked as a research engineer on industrial research projects at the Center for Education and Research on Macromolecules (CERM). He obtained his Ph.D. in 1996 in the field of polymer blends and composites under the supervision of Prof. Ph. Teyssié, Prof. R. Jérôme, and Dr. R. Fayt.

After spending 1 year as a postdoc at the Max-Planck Institute for Polymer Research in Mainz, Germany, Dr. Harrats joined Prof. G. Groeninckx's group at KULeuven. His main research areas are the material development and design of polymers, polymer blends, and composites. Dr. Harrats has conducted numerous research projects in collaboration with companies such as General Electric (USA), Alcatel-Alsthöm (France), DSM Research (Netherlands), and Redco-Eternit (Belgium).

He has joined the group of Prof. P. J. Lemstra at the Technical University of Eindhoven for a 2-year sabbatical under the Marie-Curie Research Project for Experienced Researchers.

Dr. Harrats has also actively participated in Europe–South Mediterranean Collaboration Tempus programs with Algerian universities including Ecole Nationale Supérieure de l'Enseignement Technique (ENSET) and Université des Sciences et Technologie d'Oran (USTO Oran).

1

Basics in Phase Morphologies of Multicomponent Polymer-Based Materials

1.1 Phase Morphology Development in Polymer Blends

The blending process generally starts with the components in the form of pellets of 3 mm or powder of 0.2 μm in approximate size. However, the target morphologies for the final product often require domain sizes in the range of 0.1 to 10 μm in order to achieve superior performance characteristics. This means that a size reduction of about three orders of magnitude is imposed on the initially introduced components in the mixing equipment [1]. Phase morphology development in immiscible polymer blends can be considered as the big change in dimension and shape the components undergo during their transformation and compounding operations. The minor phase in an immiscible polymer blend in the melt is deformable, as opposed to composite materials, which contain a rigid minor phase. Depending on the components characteristics and the blending conditions, a wide range of sizes and shapes (spherical, ellipsoidal, cylindrical, ribbon-like, sheet, platelet, cocontinuous, etc.) can thus be obtained for the dispersed phase during processing [2,3]. The final phase morphology in a blend is the result of a balance between deformation–disintegration phenomena and coalescence.

1.1.1 Deformation, Breakup, and Coalescence Phenomena during Melt-Blending

1.1.1.1 Basic Principles

Historically, Einstein was the first to derive an expression for the viscosity of hard spheres in a dilute suspension [4]. Later, Oldroyd [5] considered the case where the spheres are themselves liquid. Taylor extended the study to a system where the suspending medium, as well as the dispersed spheres, are Newtonian liquids [6,7]. It was observed that when the radius of the drop is great enough or the rate of distortion is high, the drop breaks up. Taylor derived the following two equations:

$$Ca = \eta_m R \dot{\gamma} / \sigma \tag{1.1}$$

$$E = Ca[(19p + 16)/(16p + 16)] \tag{1.2}$$

From the above equations it is possible to calculate the size of the largest drop that exists in a fluid undergoing distortion at any shear rate. In these equations, the governing parameters for droplet breakup are the viscosity ratio p (viscosity of the dispersed phase to that of the matrix); the type of flow (elongational, shear, combined, etc.); the capillary number Ca, which is the ratio between the deforming stress $\eta_m \dot{\gamma}$ (matrix viscosity × shear rate) imposed by the flow on the droplet; and the interfacial forces σ/R, where σ is the interfacial tension and R is the radius of the drop. If Ca is small, the interfacial forces dominate and a steady drop shape develops. Upon exceeding a critical value Ca_{crit}, the drop becomes unstable and finally breaks up. A dimensionless parameter, E (Equation 1.2), was derived which allows one to describe dispersed-particle disintegration for Newtonian systems in shear-flow fields. According to Taylor [6,7], the apparent deformation D (Equation 1.3) of the droplet has a value of 0.5, where L is the length of the particle and B is the breadth:

$$D = (L - B)/(L + B) \qquad\qquad (1.3)$$

It has been demonstrated experimentally by Taylor [6,7] that for values of p from 0.1 to 1, droplet breakup occurred at D values between 0.5 and 0.6. The expression of E in Equation 1.2 indicates that the viscosity ratio, the shear stress, the droplet diameter, and the interfacial tension are critical variables to consider in controlling particle deformation and breakup in Newtonian fluids. In that equation, however, the coalescence, which has been later found to be critical in a breakup process, has not been considered. Figure 1.1 shows a nice illustration of the process of breakup of a polymer fiber (polyamide) in a polymer matrix (polystyrene).

Grace [6] has constructed a plot of the critical capillary number as a function of the viscosity ratio, p, under two types of flow: a simple shear flow and a hyperbolic (elongational) flow field (Figure 1.2). It is shown that droplets are stable when their Ca number is below a critical value; the deformation and breakup are easier at P within a 0.25 to 1 range for shear flow, and the elongational flow field is more effective for breakup and dispersion than the shear flow. It can also be seen that at a viscosity ratio $p > 4–5$, it is not possible to break up the drop in simple shear flow.

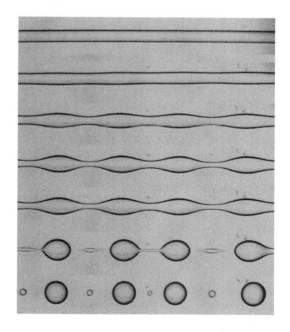

FIGURE 1.1
Sinusoidal distortions on a polyamide 6 (PA6) thread, with diameter 54 µm, embedded in a polystyrene matrix. The measurement was performed at 230°C; the times for subsequent photographs are 0 sec, 183 sec, 300 sec, 340 sec, 396 sec, 494 sec, and infinite. (From Y. Son, *Polymer* 42, 1287–1291, 2001. With permission.)

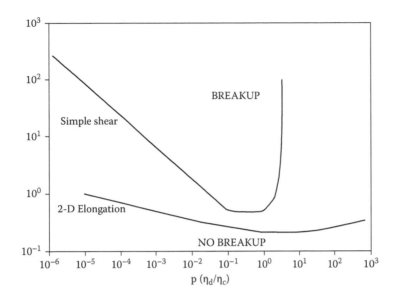

FIGURE 1.2
Critical capillary number versus viscosity ratio in simple shear and plane hyperbolic (two-dimensional elongational) flow. (From H. P. Grace, *Chem. Eng. Commun.* 14, 225, 1982. With permission.)

1.1.2   Effect of Component Characteristics on Phase Morphology

Favis [7] has summarized in a well-structured chapter the effect of the viscosity ratio, blend composition, elasticity, shear stress, and interfacial modification in immiscible blends on the final phase morphology.

1.1.2.1   *Viscosity Ratio*

In a binary blend, the ratio of the viscosity of the dispersed phase to that of the matrix is one of the most critical variables for controlling the phase morphology. In many situations, if the minor phase has a lower viscosity than the major one, the minor phase will be finely dispersed. In contrast, if the major phase exhibits a lower viscosity than the minor phase, a coarse dispersed phase will be developed.

1.1.2.2   *Blend Composition*

In a binary blend A/B, the phase morphology can be of an A-droplet-in-B-matrix when A is minor, a B-droplet-in-A-matrix when B is minor, and a third particular situation in between where the two phases A and B are cocontinuous.

As the output phase morphology in a blending operation is a balance between particle breakup and coalescence, increasing the concentration of the A component in an A/B blend will favor coalescence because of the increased particle-to-particle collision probability resulting in an increased particle size. Coalescence has been reported to start at concentration of the dispersed phase as low as 1 wt% [8]. Below a given concentration (depending on the viscoelastic characteristics of the blend constituents) the particle size gradually increases as a function of the concentration. Around that critical concentration the dimensions of the particles become very sensitive to the concentration. Whereas within a concentration window, a phase inversion region is delimited where phase cocontinuity occurs, no more particle breakup is possible and the two phases are interconnected in a tortuous structure throughout the whole blend volume. The width of this concentration window at which the phases are cocontinuous depends on the viscoelastic characteristics of the blend constituents [9].

1.1.2.2.1   Role of Coalescence

Several authors have developed empirical equations to predict the effect of coalescence on the phase morphology of binary polymer blends. Equation 1.4 relates the particle size of the dispersed phase at equilibrium to its composition in the blend [10]. This equilibrium equation was derived from a more complex expression where a rate constant for breaking the drops and one for their coalescence have been defined to account for the continuous process of phase morphology development resulting from a competition between breakup and coalescence.

$$d_e \approx \frac{24 P_r \sigma}{\pi \tau_{12}} \left(\Phi_d + \frac{4 P_r E_{DK}}{\pi \tau_{12}} \Phi_d^2 \right) \tag{1.4}$$

τ_{12} is the shear stress, σ is the interfacial tension, Φ_d is the volume fraction of the dispersed phase, E_{DK} is the bulk breaking energy, and P_r is the probability that a collision between two close particles will result in coalescence leading to phase coarsening. It is clear from the expression that the particle size at equilibrium diminishes as the shear stress increases, the interfacial tension decreases, and the volume fraction of the dispersed phase decreases. This theory has been verified experimentally for several immiscible blends where a master curve of particle size as a function of composition was found to follow a $\Phi_d + k \Phi_d^2$ dependence [11]. Fortelny et al. [12] have proposed an expression (Equation 1.5) that accounts for a drop breakup (the first term in Equation 1.5) and a coalescence process (the second term in Equation 1.5):

$$R = (\sigma_{12}(We)_c/(\eta_m \gamma)) \times (\sigma_{12} \alpha \phi/(\eta_m f_1)) \tag{1.5}$$

where R is the particle radius, $(We)_c$ is the critical Weber (or capillary) number, η_m is the viscosity of the matrix, σ_{12} is the interfacial tension, γ is the shear rate, α is the probability of coalescence after collision, ϕ is the volume fraction of the dispersed phase, and f_1 is the slope of the function $F(We)$, which describes the frequency of phase breakup at the critical Weber number.

Elmendorp and Van der Vergt also developed an expression to describe shear-induced coalescence of spherical particles [8]. A critical coalescence time has been defined which accounts for the time that passes between the arrival of a droplet at a liquid–liquid interface and the rupture of the intervening film. The critical coalescence time t_c has been defined as:

$$t_c = (3 \eta_m R/2 \sigma) \ln(R/2 h_c) \tag{1.6}$$

The above discussion clearly reveals that particle size versus composition dependence, which is affected by coalescence, is controlled to a large extent by the interfacial tension between the two phases of a binary blend. Several reports showed that the coalescence can be suppressed or at least its role minimized over a wide composition range. For example, it has been demonstrated that in a polyvinyl chloride/polyethylene (PVC/PE) blend of a low interfacial tension of 3.4 mN/m, the particle size reduction (suppressed coalescence) is much more significant than in a polyamide (PA)/PE blend having an interfacial tension as high as 14 to 18 mN/m [13–15].

The interfacial tension in immiscible polymer blends is reduced by the addition of suitable interfacial agents such as block or graft copolymers, preformed or generated *in situ* during the blending or the processing operation. The action of adding these agents to immiscible blends in an attempt to decrease their interfacial tension and reduce the coalescence phenomena during phase morphology development is called compatibilization. Literature reporting on the compatibilization of polymer blends is too extensive to be exclusively listed in this section. Addition of hydrogenated polybutadiene-b-polystyrene block copolymers as compatibilizing agents to polyethylene/polystyrene immiscible blends has been extensively investigated by Fayt et al. [16–20]. Diblock copolymers were found to be more efficient than graft, triblock, or star-shaped copolymers. In these studies the most efficient interfacial agents were reported to be diblock copolymers of a well-balanced composition. As far as the copolymer concentration is considered, stable dispersions were produced by adding as low as 1 to 2 wt% of an efficient copolymer.

1.1.2.3 Effect of Elasticity

In addition to interfacial tension, composition, and viscosity, the elasticity of the components, although not thoroughly studied, has also been considered in the understanding of the phase morphology buildup. The role of this parameter is the least understood among all parameters considered. Van Oene [21] was the first who pointed out that in capillary flow the dispersion of particles undergoes stratification and droplet–fiber formation. In addition to the role of the particle size and the interfacial tension, these two morphologies were also controlled by the differences in viscoelastic properties between the two phases. For example, in a polystyrene/polymethyl methacrylate (PS/PMMA) blend, the PMMA phase, having the largest normal stress function (more elastic than PS), exhibits a droplet morphology type in the PS matrix. After the addition of a low-molecular-weight PMMA to the blend, a stratified PMMA phase was obtained. It was also noted that when the droplet size was smaller than 1 µm, the difference in morphologies (stratified versus droplet type) disappeared. It was concluded that the elastic contribution to the interfacial tension was no longer dominant. The statement of Van Oene is that the phase of higher elasticity has the tendency to encapsulate the one with lower elasticity. He showed that it is difficult to deform a highly elastic material. He could then develop an expression where the contribution of material elasticity to the interfacial tension is considered:

$$\sigma_{eff} = \sigma + \frac{d}{12}[(N_2)_d - (N_{2m})] \tag{1.7}$$

where σ_{eff} is the effective interfacial tension under dynamic conditions, σ is the static interfacial tension, d is the droplet diameter, and N_{2d} and N_{2m} are the second normal stress functions for the dispersed phase and for the matrix, respectively. Furthermore, it has been also shown that in an extensional flow field when a Newtonian droplet is deformed by a viscoelastic matrix, no lower limiting droplet size exists beyond which disruption of particles becomes impossible [22,23]. Elmendorp and Maalcke [24] demonstrated experimentally that the normal stress exhibited by a droplet in a viscoelastic fluid stabilizes it, as predicted by

Van Oene. Levitt et al. [25] have observed that polypropylene droplets were elongated perpendicular to the flow direction in a polystyrene matrix that was highly elastic. The extent of particle stretching in the perpendicular direction of flow was found to be proportional to the normal stress differences between the phases. Particle contraction was observed upon cessation of the shearing action, which confirms the role of the elasticity on deformation.

The shear stress, $\tau_{12} = \eta_m \gamma$ ($\tau_{12} = \eta_m \gamma$), which is a machine parameter, has a direct effect on particle deformation. The particle size is inversely proportional to the applied shear stress. Increasing the shear stress results in a particle size reduction. This interrelation has been verified unambiguously in a polystyrene/polyethylene immiscible blend [26]. It has been shown that increasing the shear stress resulted in a much finer particle size and also that the viscosity ratio was less predominant compared to the shear stress. The morphology was not influenced by variations in the molecular weight of polyethylene or by an increase in the mixing temperature. Other authors reported that a variation of the shear stress by a factor of 2 or 3 does not influence significantly the particle size [27–29]. Apparently, the variation of shear stress should be much higher in order to be effective in affecting predominantly the size of the particles.

The shear stress was reported to have an effect on the shape of the dispersed phase. In a polyamide/polyethylene blend extruded through a slit die, an increase of shear stress from 17 kPa to 29 kPa resulted in a change of the nylon phase from a spherical to a fiber-like particle shape [30].

1.1.2.4   Role of the Mixing Equipment and Compounding Conditions

The mixing conditions play a crucial role in the development of phase morphology. These include the mixing equipment and the mixing parameters such as the mixing temperature, the mixing time, and the rotation speed of the mixing elements. Three major types of equipment are employed in the processing of polymer blends: internal mixers and extruders (single- and twin-screw extruders) and injection molding, which is used to shape an already compounded blend. These devices are equipped with screws or rotor blades revolving in a closed chamber or a barrel, respectively. The design and condition of the screws or the rotor blades determines the type of flow fields to which the polymer melts are subjected.

Internal mixers are mainly used for rubber-based blends, but they are also useful on a laboratory scale for blending a wide range of thermoplastics. They are practical for blending small amounts of material and allow a monitoring of torque (a viscosity-dependent parameter). Favis has reported on the influence of the mixing time on the size of the dispersed phase in a Brabender mixer [28]. By using polycarbonate/polypropylene immiscible blends, the most significant particle size reduction or phase morphology generation took place within the first 2 min of the mixing operation. Between 2 and 20 min of mixing time, very little particle size reduction of the dispersed phase was noticed. An increase in rotor speed had only a minor effect on further size reduction. Similar observations were also made by Schreiber and Olguin in polyolefin/elastomer blends prepared using internal mixer [31].

Single-screw extruders, although limited in developing high shear rates and good distributive mixing, remain, nevertheless, a widespread tool for blending polymers. Lindt and Ghosh have studied phase morphology evolution in a single-screw extruder. It was revealed that the blend undergoes a continuous change in its developing morphology during the passage through the melting zone. Within a fraction of a second, the length scale of the dispersed phase drops by several orders of magnitude [32].

The most appropriate mixing tool and the most widespread in processing polymer blends is the twin-screw extruder. Corotating and counterrotating twin-screw extruders

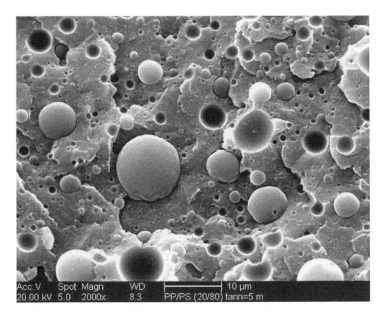

FIGURE 1.3

Typical droplet-in-matrix (or dispersed type) phase morphology in melt-blended binary blend. Scanning electron microscopy (SEM) photomicrograph of a cryofractured surface of 80 wt% polystyrene/20 wt% polypropylene melt-mixed blend. (From G. Lei, *Development of Three Phase Morphologies in Reactively Compatibilized Polyamide 6/Polypropylene/Polystyrene Ternary Blends*, master's thesis, Katholieke Universiteit Leuven, Belgium, 2004.)

are high-intensity mixing devices consisting of two screws with a kneading section for intensive mixing. An entire screw can be built using a set of kneading elements to meet specific mixing requirements. Many studies were devoted to the comparison of the mixing efficiency between a static mixer and a twin-screw extruder [33,34]; the static mixer produced a coarser dispersion than the twin-screw extruder (a factor of 2 has been reported). The dependence of phase size on the viscosity ratio for blends prepared in a twin-screw extruder was found to resemble that observed for Newtonian fluids in an elongational flow field. Detailed aspects on phase morphology development in a twin-screw extruder are well documented elsewhere [7].

1.1.3 Main Types of Phase Morphologies

There exist in polymer blends two or three major types of phase morphologies, depending on whether the encapsulated structures (composite droplets) are considered as a class apart. The most common is the droplet-in-matrix (as, for example, Figure 1.3), the (droplet-in-droplet)-in-matrix (as, for example, Figure 1.4), and the cocontinuous phase morphology where both phases are mutually interconnected throughout the whole volume of the blend (as, for example, Figures 1.5 and 1.6).

1.1.3.1 Droplet-in-Matrix Phase Morphology (Dispersed Morphology)

The mechanism and the parameters that control the droplet formation in a binary immiscible blend have been discussed in the preceding section. The size of the dispersed phase in a polymer blend is measured on images observed by microscopy. For well-mixed

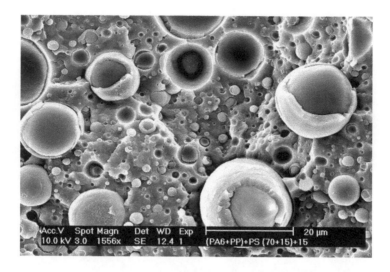

FIGURE 1.4
An illustration of a composite (encapsulated droplet-in-matrix) phase morphology in melt-blended ternary blend: 70 wt% polyamide/15 wt% polystyrene/15 wt% polypropylene. The droplet is polypropylene, the encapsulating phase is polystyrene, and the matrix is polyamide. (From G. Lei, *Development of Three Phase Morphologies in Reactively Compatibilized Polyamide 6/Polypropylene/Polystyrene Ternary Blends*, master's thesis, Katholieke Universiteit Leuven, Belgium, 2004.)

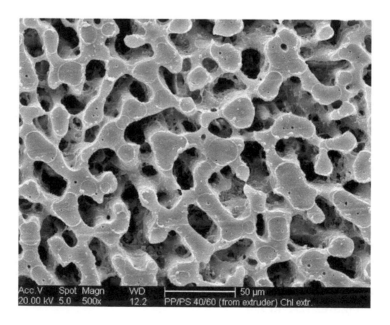

FIGURE 1.5
Typical "thick" cocontinuous phase morphology in melt-blended binary blend. A scanning electron microscopy photomicrograph of a cryosmoothed and chloroform-etched 60 wt% polystyrene/40 wt% polypropylene melt-mixed blend. The etched phase is polystyrene. (From G. Lei, *Development of Three Phase Morphologies in Reactively Compatibilized Polyamide 6/Polypropylene/Polystyrene Ternary Blends*, master's thesis, Katholieke Universiteit Leuven, Belgium, 2004.)

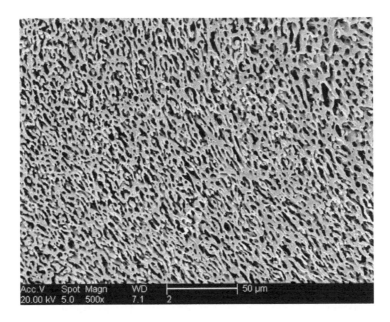

FIGURE 1.6
Typical "thin" cocontinuous phase morphology in melt-blended binary blend. A scanning electron micros-
copy photomicrograph of a cryosmoothed and chloroform-etched 60 wt% polycyclohexylmethacrylate/40 wt%
polypropylene melt-mixed blend. The etched phase is polycyclohexylmethacrylate. (From Y. Mewis, *Reactive
Compatibilization of Polymer Blends: Melt-Processing, Phase Morphology, and Properties*, master's thesis, Katholieke
Universiteit Leuven, Belgium, 2003.)

blends, the dispersed phase exhibits log-normal behavior. That is, the observed frequency
of size versus the log of the diameter of the particles results in a normal distribution.
This distribution does not appear to be substantially altered by changes in interfacial
energy or compatibilization (copolymer formation), although there may be an effect on the
mean size of the particles.

In many melt-mixing devices not all of the resin may experience the same shear history,
and bimodal or multimodal distributions may be generated. Deviation from log-normal
behavior may be provoked by reprocessing of instabilized blends or by segregation effects.
The most complete description of the dispersed phase size is contained in the frequency-
distribution histogram or the cumulative frequency histogram. It is often advantageous to
construct independent plots based on both the number fractions and the weight fractions
of particles in each group. The number fraction provides the most important information
about the particle size on the small end of the distribution curve, whereas the weight
fraction reveals the importance of larger particles. For an accurate characterization of a
droplet-in-matrix phase morphology, it is more judicious, prior to observation using scanning electron microscopy, to smoothen the surface at a temperature below the glass transition temperatures of both the matrix and the dispersion. It is also preferable to etch the
particles on the smooth surface by using a selective solvent.

Another phenomenon that has to be considered in phase-morphology development is
the transition from a droplet to a fiber-like dispersion. This transformation depends on
the extent of the deforming forces, the capillary instabilities, and the coalescence as well
as on the interfacial tension between the phases. The type of flow to which a particle is
subjected, whether it is a shear flow or an elongational flow field, is a crucial parameter to

consider when targeting a fiber-like dispersion. Elongational flow fields are more effective in inducing a droplet–fiber transition. Fibers were reported to be produced at the entrance of a capillary where elongational flow is generated [35]. Uniaxial melt drawing, which generates an elongational flow field, was also reported to generate fiber dispersions [30,36]. The coalescence of particles in a concentrated dispersed phase system where particle–particle interactions are intense can also generate fibers. This phenomenon was demonstrated mainly in the entrance zone of a capillary rheometer [37,38].

1.1.3.2 Composite Droplet Phase Morphology

The droplet-in-droplet-in-matrix or composite droplet morphology is a structure well known in high-impact polystyrene (HIPS). It is composed of polystyrene matrix and minor rubber phase that contains subinclusions of polystyrene. This morphology results from the polymerization of a mixture of 5 to 10% of polybutadiene in styrene monomer [39]. Polystyrene subinclusions are entrapped within polybutadiene particles after polymerization. Polyamide 6 was reported to be toughened by a composite dispersed phase of styrene-ethylene-butylene-styrene (SEBS) block copolymer and polyphenylene oxide (PPO) as a thermoplastic [40].

Hobbs and coworkers reported on the spontaneous development of composite droplet morphology of a series of ternary blends [41]. The work highlighted the phase morphology when three components, having by pairs different interfacial tension, were melt-blended together. The encapsulation effect responsible for the generation of composite droplets was explained in terms of the differences in interfacial tensions between each two components. It was deduced that the subinclusion formation could be predicted using Harkin's equation (Equation 1.8),

$$\lambda_{31} = \sigma_{12} + \sigma_{32} + \sigma_{13} \tag{1.8}$$

where λ_{31} is the spreading coefficient for component 3 to encapsulate component 1 and the three terms to the right are the interfacial tensions between the respective 1,2; 3,2; and 1,3 polymers. When λ_{31} is positive, encapsulation of component 3 about component 1 will take place. Compatibilization of one pair has been shown to result in composite droplet morphology, for which the third component forms subinclusions in the dispersed phase [42,43].

1.1.3.3 Cocontinuous Phase Morphology

The cocontinuous two-phase morphology consists of two coexisting, continuous, and interconnected phases throughout the whole blend volume. In contrast, in isotropic cocontinuous two-phase morphology the two blend components contribute simultaneously to the properties of the blend in all directions. Cocontinuous microphase morphologies can be frozen in from partially miscible blends during the spinodal demixing process as a result of the thermodynamic instability of the induced molecular miscibility, whereas cocontinuous micro- or macrophase morphologies can be generated via melt-blending of immiscible polymers under a particular set of conditions including the blend composition, the component characteristics, and the machine parameters selected for their melt-processing.

1.1.3.3.1 In Partially Miscible Blends

A large number of homopolymer pairs exhibit partial miscibility when mixed together either in a melt-blending process or via precipitation from a single solution. Usually, the partial miscibility is revealed only upon thermal treatment to which is subjected the molecularly

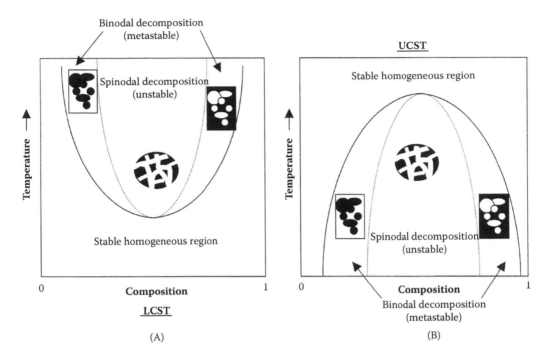

FIGURE 1.7
Scheme showing typical temperature composition, lower critical solution temperature (LCST), and upper critical solution temperature (UCST), phase diagrams, and the binodal and spinodal decomposition ranges.

miscible mixture (heating and cooling operations). A characteristic temperature–composition phase diagram is then identified for each couple of partially miscible homopolymers. As shown in Figure 1.7, an example of the typical phase diagrams of a partially miscible pair of polymers contains three distinct zones of miscibility: a stable homogeneous region, an unstable homogeneous region, and in between a two-phase metastable zone.

When the curve delimiting the homogeneous and the heterogeneous zones is convex, the mixture is identified as exhibiting lower critical solution temperature (LCST) behavior. If a miscible A/B blend having a composition slightly different from the diluted extreme cases is heated above any temperature, the LCST curve demixes into two distinct phases, A and B.

Some polymer mixtures are miscible at high temperatures but demix at lower ones as in Figure 1.7B. In this case the phase diagram of the miscibility of the blend obeys an upper critical solution temperature (UCST)—a concave curve. Indeed, a stable A/B miscible blend generated at a temperature above the UCST can demix into two A and B phases if cooled below any of the temperatures on the UCST curve. Two different main mechanisms of decomposition, a binodal and a spinodal, are known by which a partially miscible blend evolves to an immiscible state. In the meta-stable (binodal decomposition) region phase separation occurs via a nucleation and growth as in the crystallization process in semicrystalline polymers. A nucleus is formed on which are grown separated phases A and B in a droplet-in-matrix phase morphology type. Depending on the composition of the blend, an A-dispersed-in-B matrix or a B-dispersed-in-A matrix phase morphology can result. In the unstable region, the spinodal decomposition mechanism prevails. This is an unstable process that is initiated by thermal fluctuations that are already present in the miscible blend leading to demixed morphologies via spinodal decomposition. The final phase morphology resulting from this process of phase separation is a cocontinuous two-phase morphology where both A and B phases coexist in an interconnected network throughout the whole blend volume.

Cahn and Hilliard [44] were the first to develop, using the mean field approach, a theory about the dynamics of phase separation for metallurgical applications. For the spinodal decomposition, three stages having three mechanisms of domain growth have been identified including molecular diffusion, liquid flow, and phase coalescence [45]. The first stage follows the Oswald ripening law and holds for scales comprised within 2 to 9 nm up to a scale 5 times larger. The flow regime dominates when the phase size is about 1 μm. Above that scale, coalescence dominates, leading to irregular structure. As most of the investigations carried out on spinodal decomposition have focused on the effects of temperature and blend composition on phase equilibrium, effects of shear rate and pressure, which are important in industrial processes, have been neglected. For example, the LCST of a polycarbonate and polybutylene terephthalate miscible blend was increased by 60°C upon extrusion, causing miscibility. The extruded blend exhibited phase separation via spinodal decomposition, leading to a cocontinuous phase morphology that imparted to the blend excellent performance [46].

1.1.3.3.1.1 Thermoplastic/Thermoplastic Blends Polymethylmethacrylate/styrene-acrylonitrile copolymer having an acrylonitrile content within a determined range of 5.7 to 38.7 wt% is a typical example of a partially miscible blend exhibiting an LCST miscibility phase diagram that has been largely studied and reported over the years in literature [47–51]. However, many other partially miscible blends exist from which it is possible to generate cocontinuous phase morphology by controlling the cooling rate within the phase diagram and falling into the conditions of phase separation via a spinodal decomposition mechanism. A nonexclusive list includes polymethylemethacrylate/poly α-methyl-styrene-acrylonitrile, polymethylemethacrylate/chlorinated polyethylene, polymethylemethacrylate/polyvinylchloride, polyethersulphone/polyethylene oxide, styrene-butadiene rubber/butadiene rubber, and acrylonitrile-butadiene rubber/styrene-acrylonitrile blends.

1.1.3.3.1.2 Thermoplastic/Thermoset Blends Recently, there has been intensive activity on thermoplastic/thermoset blend systems where a thermoplastic polymer, initially miscible with the precursor (a curable monomer) of the thermosetting polymer, phase-separates upon a curing operation leading to a two-phase blend [52–54]. In this case the phase separation is induced by chemical reaction between the species of the initially miscible blend. Depending on the curing conditions of the base monomer of the thermoset resin (concentration of the curing agent, the curing temperature, the kinetics of curing versus phase separation), it is possible to generate cocontinuous phase morphology via the spinodal decomposition process. Kim et al. investigated the phase morphology development in a thermoset/thermoplastic blend system containing triglycidyl p-aminophenol/poly(ether sulfone) and using 4,4′-diaminophenylsulfone as the curing agent [55]. When a nonreactive poly(ether sulfone) was used, a spherical domain structure was generated. In contrast, a cocontinuous two-phase morphology was obtained when reactive poly(ether sulfone) bearing amine end-groups was employed. That clearly suggests that the spinodal decomposition rate was suppressed by the cure reaction and a delayed phase coarsening resulted from the *in situ* generation of poly(ether sulfone)-epoxy block copolymer (the phase-separated morphology was fixed at an early stage of the spinodal decomposition). A nanoscale cocontinuous phase morphology was generated. A size of the characteristic length of the cocontinuity identified as the periodic distance by the authors was about 20 nm. With the nonreactive polymer this dimension was increased from 20 nm to 1 μm.

In a mixed system (i.e., a 20/10 reactive/nonreactive polymer) the dimension can be reduced from 0.7 μm to 0.1 μm.

1.1.3.3.2 Cocontinuous Morphologies in Melt-Mixed Polymer Blends

Pötschke and Paul in their review on cocontinuous structures in immiscible melt-mixed blends gave an excellent overview of the open literature dealing with cocontinuous phase morphology as an intermediate stage in polymer blending [56]. In their attempt to understand the initial stages of phase morphology development during compounding, Sundararaj et al. [58] and Shih et al. [57,59–62] could visualize that an initial phase morphology composed of softened sheets of the major phase, which exhibit a higher melting point than that of the minor phase, are forming an interconnected structure in the continuous minor phase. As the observed sheets melt, they are progressively broken up and coalesce to form the final matrix that causes an intensive breakup of the minor phase. The ultimate and final phase morphology is composed of the major phase where are dispersed particles of the minor phase. An illustrative practical example of the formation of a cocontinuous structure as an intermediate step in the formation of blends containing a low-melting-point polyethylene and high-viscosity polystyrene has been reported [63]. A polyethylene content within as low as 0.5 to 7.8 wt% was able to form the continuous phase before the polystyrene major phase got completely softened and caused the breaking of the minor phase, leading to an inverted situation where the polyethylene became the dispersed phase.

Processing conditions or chemical reactions occurring in one or both phases of the blend can strongly affect the phase inversion. Of course, these two parameters have a direct effect on the viscosity ratio of the components. The same blend of polyamide/styrene-acrylonitrile copolymer developed phase morphology where PA6 is the matrix when processed using a single-screw extruder, whereas the inverse situation occurred when the blend was mixed several times in a laboratory mixer.

Cocontinuous phase morphology is developed in melt-mixed blends under a particular set of processing conditions, component characteristics, and blend compositions. As described above, melt-blended immiscible homopolymers A and B can exhibit a cocontinuous two-phase morphology at a specific condition involving mainly the following:

- The viscosity ratio of the components, which depends on the melting temperature at which the blend is formed, and on the shear rate, which itself depends on the mixing rpm of the mixer used.
- The blend composition (content of A and B components).
- Many of the microscopic illustrations presented in this book deal with cocontinuous phase morphologies in melt-mixed blends.

Much extended knowledge on various aspects of phase morphology, both theoretical and experimental, are published elsewhere [64].

References

1. Scott, C. E. and N. D. B. Lazo, Chap. 5 (pp. 113–139) in *Reactive Polymer Blending*, W. Baker, C. Scott, and G.-H. Hu (Eds.), Hanser, Munich (2001).
2. Utracki, L. A. and B. D. Favis, Polymer alloys and blends, in *Handbook of Polymer Science and Technology*, Vol. 4, CRC Press, Boca Raton, FL, 1989.

3. Favis, B. D., *Canadian J. Chem. Eng.*, 69, 619 (1991).
4. Einstein, A., *Ann. Physik*, 19, 289 (1906).
5. Oldroyd, J. G., *Proc. Royal Soc.*, London, A232, 567 (1955).
6. Taylor, G. I., *Proc. Royal. Soc.*, London, A138, 41–48 (1932).
7. Taylor, G. I., *Proc. Royal Soc.*, London, A146, 501–523 (1934).
8. Elmendorp, J. J. and A. K. Van der Vergt, *Polym. Eng Sci.*, 26, 1332 (1986).
9. Harrats, C. and N. Mekhilef, Co-continuous phase morphologies: predictions, generation, and practical applications, in *Micro- and Nanostructured Multiphase Polymer Blend Systems: Phase Morphology and Interfaces*, C. Harrats, S. Thomas, and G. Groeninckx (Eds.), Taylor & Francis, CRC Press, New York, 2006.
10. Tokita, N., *Rubber Chem. Technol.*, 50, 292 (1977).
11. Favis, B. D. and J. M. Willis, *J. Polym. Sci., Polym. Phys.*, 28, 2259 (1990).
12. Fortelny, I., Z. Cerna, J. Binko, and J. Kovar, *J. Appl. Polym. Sci.*, 48, 1731 (1993).
13. Liang, H., B. D. Favis, Y. S. Yu, and A. Eisenberg, *Macromolecules*, 32, 1637 (1999).
14. Chapleau, N., B. D. Favis, and P. J. Carreau, *J. Polym. Phys.*, 36, 1947 (1998).
15. Willis, M., V. Caldas, and B. D. Favis, *J. Mater. Sci.*, 26, 4742 (1991).
16. Fayt, R., R. Jérôme, and Ph. Teyssié, *Makromol. Chem.*, 187, 837 (1986).
17. Fayt, R., R. Jérôme, and Ph. Teyssié, *J. Polym. Sci., Polym. Lett.*, 24, 25 (1986).
18. Fayt, R., R. Jérôme, and Ph. Teyssié, *Polym. Eng. Sci.*, 27, 328 (1987).
19. Fayt, R., R. Jérôme, and Ph. Teyssié, *J. Polym. Sci., Polym. Phys.*, 27, 775 (1989).
20. Fayt, R., C. Harrats, S. Blacher, R. Jérôme, and Ph. Teyssié, *J. Polym. Sci., Polym. Phys.*, 69, 178 (1993).
21. Van Oene, H., *J. Colloid. Inter. Sci.*, 40, 448 (1972).
22. Flumerfelt, R. W., *Ind. Eng. Chem. Fundam.*, 11, 312 (1972).
23. Janssen, M. H. and H. E. H. Meijer, *J. Rheol.*, 37, 597 (1993).
24. Elmendorp, J. J. and R. J. Maalcke, *Polym. Eng. Sci.*, 25, 1041 (1985).
25. Levitt, L., C. W. Macosko, and S. D. Pearson, *Polym. Eng. Sci.*, 36, 1647 (1996).
26. Min, K., J. L. White, and J. F. Fellers, *J. Appl. Polym. Sci.*, 29, 2117 (1984).
27. Favis, B. D. and J. P. Chalifoux, *Polym. Eng. Sci.*, 27, 1591 (1987).
28. Favis, B. D., *J. Appl. Polym. Sci.*, 39, 285 (1990).
29. Sandararaj, U. and C. W. Macosko, *Macromolecules*, 28, 2647 (1995).
30. Gonzalez-Nunez, R., D. DeKee, and B. D. Favis, *Polymer*, 37, 4689 (1996).
31. Schreiber, H. P. and A. Olguin, *Polym. Eng. Sci.*, 23, 129 (1983).
32. Lindt, J. T. and A. K. Ghosh, *Polym. Eng. Sci.*, 32, 1802 (1992).
33. Favis, B. D. and D. Therrien, *Polymer*, 32, 1474 (1991).
34. Plochocki, P., S. S. Dagli, J. Starita, and J. E. Curry, *J. Elastomers. Plast.*, 18, 256 (1986).
35. Dreval, V. E., G. V. Vinogradov, E. P. Plotnikova, M. P. Zabugina, N. P. Plotnikova, E. V. Kotova and Z. Pelzbauer, *Rheol. Acta*, 22, 102 (1983).
36. Gonzalez-Nunez, R., C. F. Chan Man Fong, B. D. Favis, and D. DeKee, *J. Appl. Polym. Sci.*, 62, 1627 (1996).
37. Tsebrenko, V., A. V. Yudin, T. I. Ablazova, and G. V. Vinogradov, *Polymer*, 17, 831 (1976).
38. Tsebrenko, V., G. P. Danilova, and A. Y. Malkin, *J. Non-Newt. Fl. Mech.*, 31, 1 (1989).
39. Amos, J. L., O. R. McIntyre, and J. L. McCurdy, U.S. Patent 2,694,692 (1954).
40. Campbell, J. R., S. Y. Hobbs, T. J. Shea, and V. H. Watkins, *Polym. Eng. Sci.*, 30, 1056 (1990).
41. Hobbs, S. H., M. E. Dekkers, and V. H. Watkins, *Polymer*, 29, 1598 (1988).
42. Legros, A., P. J. Carreau, B. D. Favis, and A. Michel, *Polymer*, 38, 5085 (1997).
43. Guo, H. F., S. Packirisamy, N. V. Gvozdic, and D. J. Meier, *Polymer*, 38, 785 (1997).
44. Cahn, J. W. and J. E. Hilliard, *J. Chem. Phys.*, 31, 688 (1958).
45. Siggia, E. D., *Phys. Rev. A*, 20, 595–605 (1979).
46. Utracki, L. A., *Polymer Blends Handbook*, Utracki, L. A. (Ed.), Kluwer Academic Publishers, Dordrecht, 2002.
47. Kessler, J., H. W. Kammer, and K. Klostermann, *Polym. Bull.*, 15, 113 (1986).
48. Macbrierty, V. J., D. C. Douglass, and T. K. Kwei, *Macromolecules*, 11, 1265 (1978).

49. Cho, K., J. Kressler, and T. Inoue, *Polymer*, 35, 1332 (1994).
50. Higashida, N., J. Kressler, and T. Inoue, *Polymer*, 36, 2761 (1995).
51. McMaster, L. P. and O. Olabisi, *ACS Org. Coat. Plast. Chem. Prepr.*, 35, 322 (1975).
52. Yamanaka, K. and T. Inoue, *Polymer*, 30, 662 (1989).
53. Yamanaka, K., Y. Takagi, and T. Inoue, *Polymer*, 30, 1839 (1989).
54. Yamanaka, K. and T. Inoue, *J. Mater. Sci.*, 25, 241 (1990).
55. Kim, B. S., T. Chiba, and T. Inoue, *Polymer*, 36, 43 (1995).
56. Pötschke P. and D. R. Paul, *J. Macromol. Sci.: Part C: Polym. Rev.*, 43, 87 (2003).
57. Shih, C. K., *Polym. Eng. Sci.*, 35, 1688–1694 (1995).
58. Sundararaj, U. et al., *Polym. Eng. Sci.*, 36, 1769–1781 (1996).
59. Shih, C. K., Fundamentals of polymer compounding: Part II: simulation of polymer compounding process, 49th ANTEC, Montreal, SPE, 99 (1991).
60. Shih, C. K. et al., Rheological properties of multicomponent polymer system undergoing melting or softening during compounding, 48th ANTEC, Dallas, SPE, 951 (1990).
61. Shih, C. K., Advances *Polym. Techn.*, 11, 223 (1992).
62. Shih, C. K. and D. G. Tynan, *Polym. Eng. Sci.*, 31, 1670 (1991).
63. Scott, C. E. and S. K. Joung, *Polym. Eng. Sci.*, 36, 1666 (1996).
64. Harrats, C., *Micro- and Nanostructured Multiphase Polymer Blend Systems: Phase Morphology and Interfaces*, S. Thomas and G. Groeninckx (Eds.), Taylor & Francis/CRC Press, NY, 2006.

2

Phase Morphology Investigation: Microscopic Tools, Tips, and Selected Scanning Electron Photomicrographs

2.1 Introduction

It is incoherent to write about observation of phase morphology in material science without a brief introduction of the concept and development of microscopic tools. Indeed, since the discovery of optical lenses, man's desire to observe objects too small to be seen by human eyes has intensified. Although indirect physical and chemical processes and tools were in most cases enough to gain knowledge, the scientist always sought to observe these phenomena in their real, direct, and natural states and environment. Bacteria and micron-scale living or nonliving bodies were the first objects to be observed using microscopic tools.

Credit for the first microscope is usually given to Zacharias Janssen around the year 1595. Because Zacharias was very young at that time, it is possible that his father, Hans, made the first one, but young Zach took over the introduction of these seemingly magic observation tools.

The first compound microscope introduced by the Janssens was simply a tube with lenses at each end. The magnification of these early scopes ranged from 3× to 9×, depending on the size of the diaphragm openings.

In the 17th century, Hooke's remarkable engineering abilities enabled him to invent and improve many mechanical devices, including timepieces (for which he invented the spiral spring), the quadrant, and the Gregorian telescope. Perhaps even more intriguing than his actual inventions are the devices he designed but never built: he anticipated the invention of the steam engine, and as early as 1684 he described a working telegraph system.

Hooke balanced his inventions with more pure research. He improved early compound microscopes around 1660. In *Micrographia* (1665), he coined the word *cell* to describe the features of plant tissue (cork from the bark of an oak tree) he was able to discover under the microscope. He applied his extensive mathematical knowledge in formulating the theory of planetary movement, which provided a basis for Sir Isaac Newton's theories of gravitation. In 1667 he discovered the role of oxygenation in the respiratory system.

But also Antoni van Leeuwenhoek was a man with many talents; his most important attributes were creativity, power of observation, and ingenuity. Leeuwenhoek was a common man without any fortune or formal education, so he had to work for a living. Leeuwenhoek made simple (one-lens) microscopes. He was not the first person to build a microscope, but the microscopes that he could build were the best ones of that time period. Leeuwenhoek was the first person to describe bacteria (from teeth scrapings) and protozoan (from pond water) and also helped to prove the theory of blood circulation. He gained much of his inspiration from reading Hooke's *Micrographia*.

The light microscope made it possible for humankind to study the fine structure of matter. The introduction of lenses corrected for the intrinsic optical defects experienced in earlier systems and brought the light microscope to an optimal use and perfection. In spite of the introduction of polarization, dark-field, and phase-contrast microscopy, the fundamental nature of light imposes a scale limit to the use of the light microscope. An original and spectacular leap forward in the development of microscopy has been provoked by the physicists and mathematicians of the 20th century. The concept can be traced back to the second half of the 19th century, but the real theoretical tools to build an electron microscope were realized by Broglie (1924), who revealed that a moving electron can be assigned a very short wavelength, and by Busch (1926), who discovered that a suitably shaped magnetic or electrostatic field could be used as true lenses for an electron beam so as to produce a faithful and even an enlarged image. A milestone in the historical development as well as the physics and electronics involved in electron microscopes was achieved by Wischnitzer [1] in 1962.

Among the few modes used in electron microscopy, the most commonly employed is the scanning electron microscope (SEM). It is simple to use, does not require fastidious sample preparation, and straightforwardly produces an image of the surface of the object under observation. However, some limitations have to be mentioned—for example, the attainable magnification, which does not exceed 30× when good-quality images, particularly in organic materials such as polymers, are searched for. The SEM reflects the topography of the sample—that is, the top surface of the sample—whereas, in the transmission mode, information about the phase morphology of the sample is gained from the bulk. That implies the use of thin slices of the material to investigate, the thickness of which is within a few nanometers to a few microns.

Michler [2] has nicely summarized and realized practical examples of some of the modern tools of microscopy used to study the morphology and microstructure of polymers and polymer-based materials. The most frequently employed microscopic tool remains the scanning electron microscope. It is the fastest and allows one to reach interesting dimensions in multicomponent polymer blends and composites. Transmission electron microscopy can be ranked in the second position, whereas the optical microscope is usually used as a "first-check tool" before deeper investigation. It is nevertheless the strategic tool employed in life science (biomedical, biologics, etc.). In all these cases the sample preparation step is crucial before investigating the material's microstructure.

2.2   Sample Preparation for Scanning Electron Micrography

Poor preparation and weak mounting will lead to poor-quality photomicrographs due to poor electrical conduction from the specimen to the holder. Conducting specimens do not require surface metalation, but a tight adhesion to a metallic holder using conductive glue is necessary. Of course the glue should not outgas under the high vacuum applied in the microscope prior to observation. Some other techniques apply a conducting double-sided adhesive. The nonconducting materials, like most of the polymers, need to be coated using a metal including silver, gold or gold-palladium, or carbon to make their outer surfaces conductive. Care should be taken to have a uniform layer so that the surface of the visible coating is as nearly as possible a positive replica of the underlying surface of the specimen under investigation. Nowadays modern instruments exist which allow for accurate and uniform vacuum deposition of metal layers on specimens of any shapes.

Gold is the metal most frequently deposited on organic nonconducting surfaces, as is the case with most of the commercial polymers. Various preparation techniques are followed for a judicious investigation of polymer-based materials including cryosurfaces and smoothed surfaces.

Cryosurfaces are obtained by breaking a sample below the lowest glass transition temperature of the blend components. For practical reasons and to ensure sufficient and uniform cooling, liquid nitrogen is used for materials with glass transition temperatures below room temperature. This cooling medium is used for efficiency and time-saving criteria. Otherwise, the sample can be fractured below the lowest glass transition temperature of the sample component in case of multicomponent-based materials. Of course, to attain that temperature without having a risk of local and bulk warming, more time is required than with liquid nitrogen. A lack of sample cooling can be revealed only during the SEM observation. The surfaces would not be flat, but partially elongated structures can be visible due to plastic deformation. Also a profile of surface texture can be depicted from the outer surface to the internal bulk area where the sample has not been cooled sufficiently. Figures 2.1 and 2.2 are obtained from the same sample, but the surfaces have been prepared at differing cooling extent. The apparent two-phase morphologies observed are quite different, although the intrinsic phase morphologies constituting the samples are identical. The fracture surface of the sample in Figure 2.2 is smooth and the breaking is cleaner compared to Figure 2.1, which depicts deformation due to lack of cooling. The particles in Figure 2.1 are less spherical than those in Figure 2.2. The particles have been compressed and deformed significantly into elongated shapes. This can be considered as a destructive microscopic investigation test, as the intrinsic structure on the surface has been substantially altered

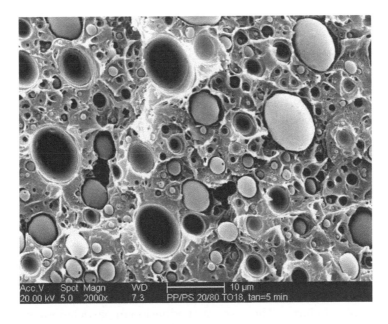

FIGURE 2.1
SEM photomicrograph of a melt-blended 20 wt% of polypropylene/80 wt% polystyrene; cryofracture surface obtained by breaking the sample in liquid nitrogen. Sample insufficiently cooled before breaking. (From T. S. Omonov, *Crucial Aspects of Phase Morphology Generation and Stabilization in Two- and Three-Phase Polymer Blends: Physical, Reactive and Combined Routes of Compatibilization*, Ph.D. thesis, Katholieke Universiteit Leuven, Belgium, 2007, under the supervision of C. Harrats and G. Groeninckx.)

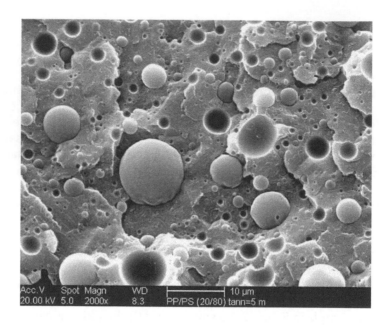

FIGURE 2.2
SEM photomicrograph of a melt-blended 20 wt% of polypropylene/80 wt% polystyrene; cryofracture sur-
face obtained by breaking the sample in liquid nitrogen: sample sufficiently cooled down before breaking.
(From T. S. Omonov, *Crucial Aspects of Phase Morphology Generation and Stabilization in Two- and Three-Phase
Polymer Blends: Physical, Reactive and Combined Routes of Compatibilization,* Ph.D. thesis, Katholieke Universiteit
Leuven, Belgium, 2007, under the supervision of C. Harrats and G. Groeninckx.)

via drawing and material flow. A perfect cryofracturing should be able to leave the phase
morphology intact as in the original material before sample preparation.

Other illustrations of the cooling artifact are shown in Figures 2.3 and 2.4. Figure 2.3
shows a blend containing 90 wt% of polypropylene (PP) in which are dispersed, in the
form of droplets, 10 wt% of polycyclohexylmethacrylate. The fracture surface of the sample in Figure 2.3 was obtained at room temperature (i.e., well above the glass transition of
polypropylene). As a direct consequence, the surface of the PP has been extensive in the
neighborhood of the particles. A substantial plastic deformation occurred upon the action
of sample breaking, resulting in misleading information about the real phase morphology
of the sample under investigation.

In the cryofracture surface, shown in Figure 2.4, no such process is visible. The cut is neat
and the matrix behaved as a glassy and perfectly brittle material. Almost no sign of plastic
deformation is visible on the PP matrix.

Another aspect to consider in investigating the phase morphology of polymer blends is
the extent of adhesion between the two components (particles and the matrix). Although
only qualitative, scanning electron microscopy can be a sufficient tool to probe the adhesion between the blend phases. The absence of adhesion in particle-in-matrix morphology
is confirmed when upon the specimen fracture the particles are detached (removed) from
the matrix, usually leaving empty cavities. They can also remain on the observed surface
as isolated particles as in Figure 2.1 through Figure 2.4. In the absence of adhesion, the
surfaces of the particles are smooth. In the case of good adhesion, however, the particles
remain cohesive to the matrix and the fracture of the sample occurs both in the matrix and

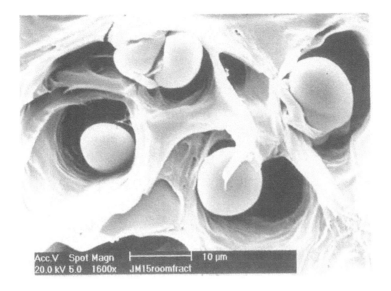

FIGURE 2.3
SEM photomicrograph of a melt-blended uncompatibilized 90 wt% polypropylene/10 wt% polycyclo-hexylmethacrylate blend; fracture surface obtained at room temperature. (From C. Harrats, T. S. Omonov, G. Groeninckx, and P. Moldenaers, *Polymer* 45, 8115, 2004. With permission.)

in the particles. In the case where some of the particles are detached from the matrix, they exhibit rough outer surfaces. These aspects are illustrated in Figure 2.5 and Figure 2.6.

Many other situations are illustrated throughout the series of images presented in this chapter. Each image is briefly described, and the important phase morphology features it reveals are also commented on. Note that it is not the objective of this book to discuss

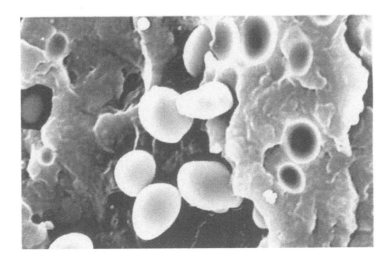

FIGURE 2.4
SEM photomicrograph of a melt-blended uncompatibilized 90 wt% polypropylene/10 wt% polycyclohexyl meth-acrylate blend; cryofracture surface obtained in liquid nitrogen. (From C. Harrats, T. S. Omonov, G. Groeninckx, and P. Moldenaers, *Polymer* 45, 8115, 2004. With permission.)

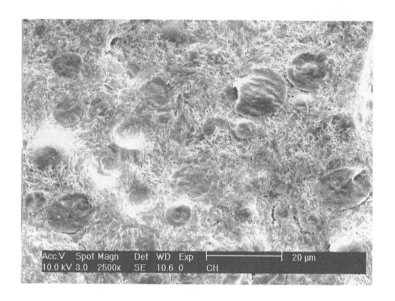

FIGURE 2.5

SEM photomicrograph of a melt-blended 30 wt% polycaprolactone/70 wt% polyethylene oxide, showing the aspects of phase morphology when the adhesion between the matrix and particles is good. Magnification: 2500×. The minor phase particles are attached firmly to the matrix. They are broken instead of being dislocated and extracted from the matrix. (From T. S. Omonov, *Crucial Aspects of Phase Morphology Generation and Stabilization in Two- and Three-Phase Polymer Blends: Physical, Reactive and Combined Routes of Compatibilization*, Ph.D. thesis, Katholieke Universiteit Leuven, Belgium, 2007, under the supervision of C. Harrats and G. Groeninckx.)

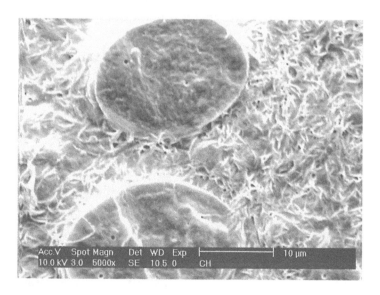

FIGURE 2.6

SEM photomicrograph showing the aspects of phase morphology when the adhesion between the matrix and particles is good; melt-blended 30 wt% polycaprolactone/70 wt% polyethylene oxide. Image magnification: 5000×. The minor phase particles are attached firmly to the host matrix. They are broken instead of being dislocated and extracted from the matrix. (C. Harrats, 2006, unpublished data.)

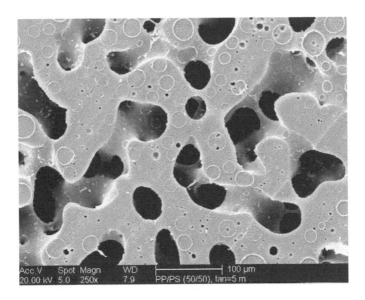

FIGURE 2.7
SEM photomicrograph (magnification: 250×, cryosmoothed surface and chloroform etched) of a melt-blended 50 wt% polypropylene/50 wt% polystyrene (PS) showing artifacts on the observed surface. When the extractable polystyrene phase is not fully and completely etched, the PS/solvent solution dries and leaves PS particles deposited on the surface of observation. This can lead to wrong characterization. As shown, the small aggregates in the big cavities are residue of PS phase not extracted with solvent-flowing process out of the matrix. Note also that the drying has generated the visible polymer rings on the surface of observation. (From T. S. Omonov, *Crucial Aspects of Phase Morphology Generation and Stabilization in Two- and Three-Phase Polymer Blends: Physical, Reactive and Combined Routes of Compatibilization,* Ph.D. thesis, Katholieke Universiteit Leuven, Belgium, 2007, under the supervision of C. Harrats and G. Groeninckx.)

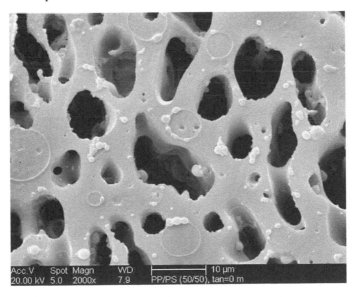

FIGURE 2.8
SEM photomicrograph (higher magnification 2000×) of the same sample as in Figure 2.6. (From T. S. Omonov, *Crucial Aspects of Phase Morphology Generation and Stabilization in Two- and Three-Phase Polymer Blends: Physical, Reactive and Combined Routes of Compatibilization,* Ph.D. thesis, Katholieke Universiteit Leuven, Belgium, 2007, under the supervision of C. Harrats and G. Groeninckx.)

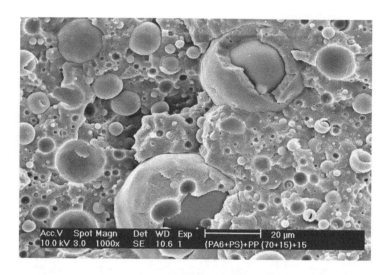

FIGURE 2.9

SEM photomicrograph (magnification 1000×) of cryofracture surfaces showing the absence of adhesion in melt-blended 70 wt% polyamide 6/15 wt% polystyrene/15 wt% polypropylene ternary blend. All the visible particles are detached from the matrix of polyamide 6 as an indication of absence of interfacial adhesion between the minor phases (PS, PP) and the host matrix polyamide 6. (From T. S. Omonov, *Crucial Aspects of Phase Morphology Generation and Stabilization in Two- and Three-Phase Polymer Blends: Physical, Reactive and Combined Routes of Compatibilization*, Ph.D. thesis, Katholieke Universiteit Leuven, Belgium, 2007, under the supervision of C. Harrats and G. Groeninckx.)

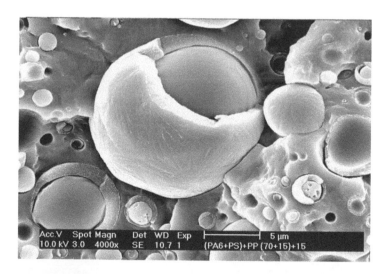

FIGURE 2.10

Higher-magnification image (4000×) of the same sample as in Figure 2.9. The composite big particle has a weak adhesion with the polyamide matrix as well as a weak adhesion with the subincluded polypropylene it encapsulates. The surfaces are very clean, indicating absence of adhesion. (From T. S. Omonov, *Crucial Aspects of Phase Morphology Generation and Stabilization in Two- and Three-Phase Polymer Blends: Physical, Reactive and Combined Routes of Compatibilization*, Ph.D. thesis, Katholieke Universiteit Leuven, Belgium, 2007, under the supervision of C. Harrats and G. Groeninckx.)

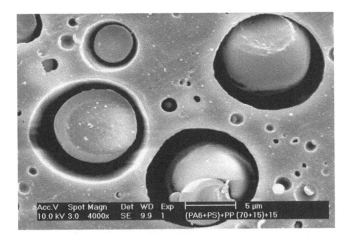

FIGURE 2.11
SEM photomicrographs (cryosmoothed surface using diamond knife; magnification 4000x) of melt-blended 70 wt% polyamide 6/15 wt% polystyrene/15 wt% polypropylene ternary blend. This sample is the same as in Figure 2.10. This image illustrates the beneficial role of the selective extraction (etching) of one of the phases in characterizing the phase morphology. The matrix is polyamide 6 and the visible spherical particles are those of polypropylene, whereas the empty voids around the particles are the spaces left after the selective removal of polystyrene using chloroform. The image reveals also that some polystyrene individual particles were dispersed in the polyamide matrix without encapsulating the polypropylene particles. This information was not possible without the selective etching of the PS phase as in Figure 2.9 and Figure 2.10, where the sample was cryofractured without etching. Quantitative information about the size of the phases is more accurate in this case, where images providing sufficient statistics are considered. The thickness of the PS shell surrounding the PP can be precisely determined using these SEM photomicrographs. (From T. S. Omonov, *Crucial Aspects of Phase Morphology Generation and Stabilization in Two- and Three-Phase Polymer Blends: Physical, Reactive and Combined Routes of Compatibilization*, Ph.D. thesis, Katholieke Universiteit Leuven, Belgium, 2007, under the supervision of C. Harrats and G. Groeninckx.)

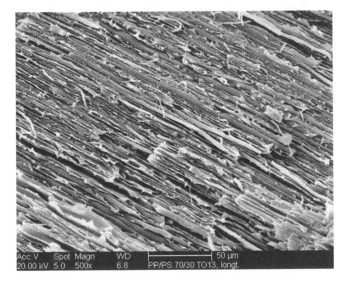

FIGURE 2.12
SEM photomicrograph of melt-blended 70 PP/30 PS blends. Polystyrene is the minor phase. It has the same cylinder-like oriented structure as in 70 PS/30 PP as a result of extensive extrusion-induced orientation in the die. (From T. S. Omonov, *Crucial Aspects of Phase Morphology Generation and Stabilization in Two- and Three-Phase Polymer Blends: Physical, Reactive and Combined Routes of Compatibilization*, Ph.D. thesis, Katholieke Universiteit Leuven, Belgium, 2007, under the supervision of C. Harrats and G. Groeninckx.)

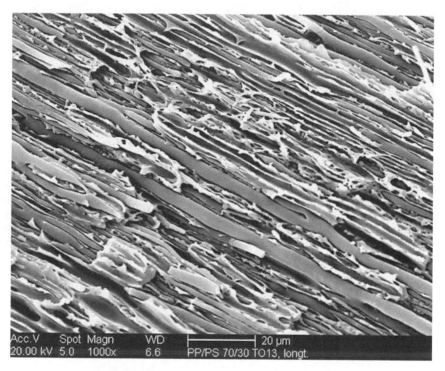

FIGURE 2.13

SEM photomicrograph of cryofractured surfaces of melt-blended polypropylene/polystyrene showing highly oriented structure. This is an additional example of anisotropic phase morphology in 70 wt% polypropylene/ 30 wt% polystyrene blends. The elongated entities of the minor phase (30 wt% polystyrene) are clearly visible in the matrix (70 wt% polypropylene). They are cylinders of infinite length as a result of extensive orientation in the die entrance during the extrusion process. (1000×magnification). (From T. S. Omonov, *Crucial Aspects of Phase Morphology Generation and Stabilization in Two- and Three-Phase Polymer Blends: Physical, Reactive and Combined Routes of Compatibilization*, Ph.D. thesis, Katholieke Universiteit Leuven, Belgium, 2007, under the supervision of C. Harrats and G. Groeninckx.)

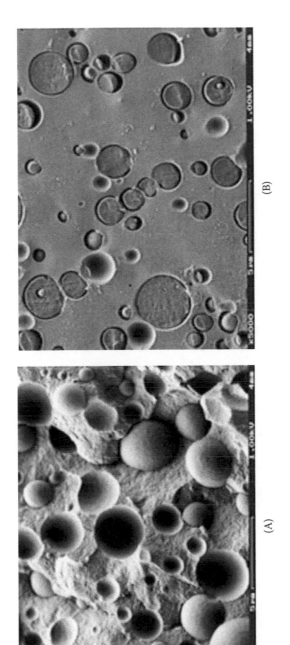

FIGURE 2.14

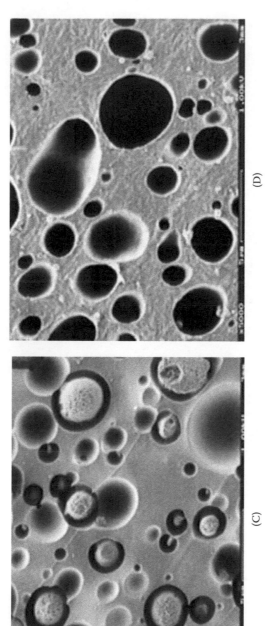

FIGURE 2.14
(Continued)

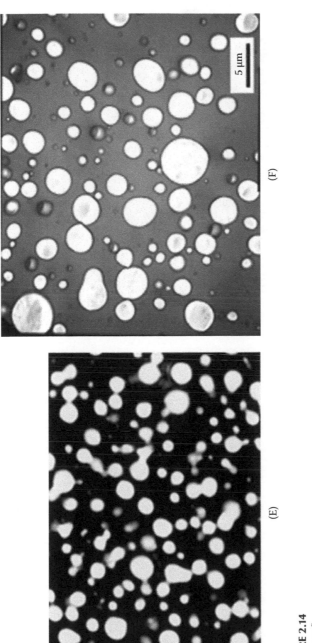

(F)

(E)

FIGURE 2.14
(*Continued*)

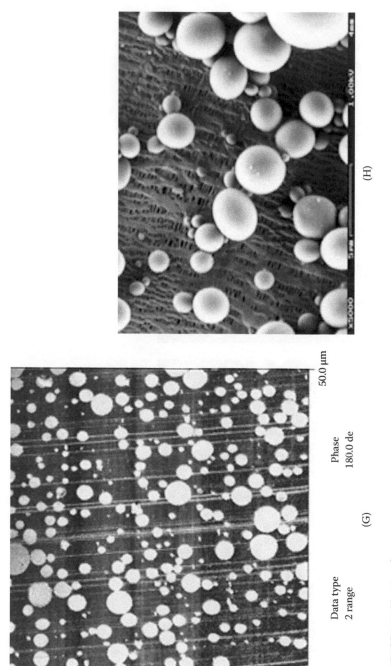

FIGURE 2.14 (*Continued*)

SEM photomicrographs. Comparison of different preparation methods for quantitative image analysis of particle sizes in matrix-particle morphologies of immiscible blends: example of thermoplastic polyurethane (TPU) with 20 wt% PP, melt-mixed using a ZSK-30 extruder. For sample preparation extruded strands or granules were taken and differently treated. All images show the same sample: (A) cryofractured, (B) cryomicrotomed, (C) plasma-etched surface, (D) etched surfaces using boiling xylene, (E) light microscopy, (F) transmission electron microscopy (TEM) photomicrograph, (G) atomic force microscopy (AFM), (H) TPU dissolved and the dispersed particles separated. (From P. Pötschke, K. Wallheinke, A. Janke, C. Bellmann, H. Stutz, and W. Heckmann, *J. Macromol. Sci. Phys. B* 38, 527–540, 1999. With permission.)

deeply the phase morphology features in each of the images. Most of these photomicrographs are published elsewhere in different forms. This book provides the opportunity to show the versatile aspects that scanning and transmission electron microscopy allow observing on the phase morphology of various polymer blend and nanocomposite systems. Some typical phase morphologies representative of the huge number of photomicrographs published in open literature are shown as a selection. A particular emphasis is put in this book on phase morphologies in nanocomposites and also on features of the cocontinuous phase morphology in binary polymer blends.

References

1. Wischnitzer, S., *Introduction to Electron Microscopy*, Pergamon Press, New York, 1962.
2. Michler, G. H., *Ultramicroscopy*, 15, 81–100 (1984).

3

Selected Microscopic Illustrations of Phase Morphology in Nanocomposites

3.1   Introduction

Polymer/layered silicate clay nanocomposites have attracted a great deal of interest in both industry and academia. The starting point was the first successful revelation of the intercalation/exfoliation concept by Toyota researchers in Japan in the early 1990s on a nylon/clay combination [1–5]. The main reasons behind the intense research activity in this field are the enhanced properties of nanocomposites compared to the pure polymers and the conventional fiber- and filler-reinforced composites. They exhibit a large increase in elasticity modulus, strength [6–8], and heat resistance [9], in addition to a substantial decrease in gas permeability [10–12] and flammability [12–14]. Polymer nanocomposites are very valuable when used in coating, structural, and packaging materials in a wide range of applications. It has to be emphasized that these properties require full exfoliation of the clay into very individual platelets, uniformly distributed throughout the nanocomposite volume.

After a slowdown of the activity due to the use of organic solvents to achieve exfoliation of the clay, the research success of Vaia et al. [15] in solvent-free melt exfoliation of montmorillonite (MMT) clay directly in extruders led to a new boost in scientific and industrial competitions.

The present introduction is not intended to review the huge amount of literature generated each year on nanocomposites. The reader interested in the full area is advised to consult the extensive review of the progress achieved in the field of nanocomposites till 2003, which has been published by Suprakas and Okamoto [16].

3.2   Some Specific Aspects Related to the Preparation of Polymer Nanocomposites

A period of great optimism followed the discovery, first published by Toyota researchers [4,18], of the concept of exfoliating clays in polymer matrices. Indeed, the produced nanocomposites have faced challenges in two distinct and crucial stages. One concerns the various applications viewed of a material made of a combination of as low as 1 to 4 vol.% of clay in a polymer exhibiting stiffness and barrier properties in some cases much higher than a classical composite containing more than 20 vol.% of reinforcement. The optimism has quickly been moderated by the limitations encountered in using the most interesting mass polymers such as polyolefins. Contrary to polar polymers such as polyamides and few polyesters in which clay was exfoliated by a simple modification of the classical

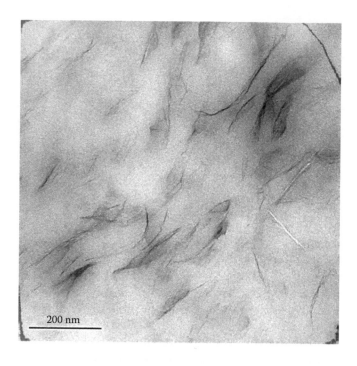

FIGURE 3.1

Transmission electron microscopy (TEM) photomicrograph of ethylene vinyl acetate nanocomposites containing 5 wt% of unmodified cloisite organoclay C20A prepared using a melt-compounding process at a temperature of 170°C, a screw rotation of 200 rpm, for a mixing time of 10 min. (From P. Nawani, P. Desai, M. Lundwall, M. Y. Gelfer, B. S. Hsiao, M. Rafailovich, A. Frenkel, A. H. Tsou, J. W. Gilman, and S. Khalid, *Polymer* 48, 827–840, 2007. With permission.)

compounding methods (water-assisted extrusion for clay/polyamide), or by a simple chemistry operated on the clay to make it more polar and able to exfoliate, the hydrophobic polymer matrices require a much more complex technique.

This situation can be considered as the starting point for fastidious and complex research activities launched in many polymer laboratories to develop adequate and relatively cheap methods to exfoliate clay in hydrophobic polymer matrices [18–24].

Three main categories of techniques of producing nanocomposites are well known nowadays: through polymerization of the monomer from the surfaces of the clay, which is intercalated in the polymerization medium via melt-compounding; where the polymer and the modified clay are mixed together using extruders; or by using a solvent in which the polymer is soluble and the clay is swollen or fully exfoliated. The following are some examples of nanocomposites prepared via the three different techniques:

- Polyamide 6 [25,26], epoxy [27,28], poly(methylmethacrylate) [29,30], poly(ε-capro-lactone) [31], polyurethane [32], polyimide [33,34], and polyamide 12 [35] were prepared via polymerization pathway.

- Polyester [36,37], polyamide 6 [38,17], polypropylene [39,40], polyethylene [41], and polyamide 66 [42] were prepared via a melt-compounding process.

- Polypropylene [43], polyethylene [44], poly(vinyl-pyrrolidon) [45], poly(vinyl alcohol) [46], and polyethylene oxide [1] were produced via solution blending.

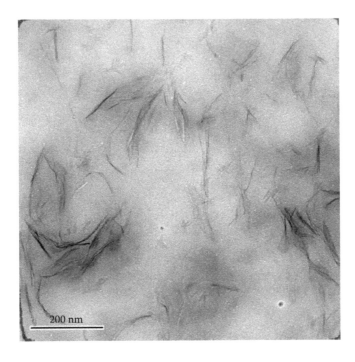

FIGURE 3.2
TEM photomicrograph of ethylene vinyl acetate nanocomposites containing 5 wt% of cupper-modified organoclay cloisite C20A, prepared using a melt-compounding process at a temperature of 170°C, 200 rpm screw rotation, for a mixing time of 10 min. Modified organoclays were prepared by treating the organoclay suspensions in alcohol (ethanol or methanol) with a transition metal ion (cupper) solution in the same solvent. (From P. Nawani, P. Desai, M. Lundwall, M. Y. Gelfer, B. S. Hsiao, M. Rafailovich, A. Frenkel, A. H. Tsou, J. W. Gilman, and S. Khalid, *Polymer* 48, 827–840, 2007. With permission.)

Besides the intense efforts devoted to the development of exfoliation pathways, extensive research activity is dealing with the understanding of the underlying fundamental aspects associated with the clay dispersion, the resulting physical properties, or the interrelationship between both. The improved properties of the nanocomposite are ascribed either to the polymer–clay interface [41,47] or to the clay aspect ratio [6,33]. According to Shelley et al. [48], the enhancement of stiffness, strength, and barrier properties is the consequence of the existence of a confined polymer fraction having a higher local stiffness.

The most crucial fundamental aspect, however, in studying nanocomposites resides in the insertion (intercalation) of large polymer molecules into the gallery space of the clay.

3.3 Mobility of Polymer Molecules in Confined Spaces of Layered Silicates

The properties of polymer/clay nanocomposites depend largely on the dispersion of the silicate layers, the interaction between the polymer molecules, and the silicate surface, which in turn depends on the type of surfactants that are used.

FIGURE 3.3
TEM photomicrograph of ethylene vinyl acetate nanocomposites containing 30 wt% of unmodified cloisite organoclay C20A prepared using a melt-compounding process at a temperature of 170°C, a screw rotation of 200 rpm, for a mixing time of 10 min. Modified organoclays were prepared by treating the organoclay suspensions in alcohol (ethanol or methanol) with a transition metal ion (cupper) solution in the same solvent. (From P. Nawani, P. Desai, M. Lundwall, M. Y. Gelfer, B. S. Hsiao, M. Rafailovich, A. Frenkel, A. H. Tsou, J. W. Gilman, and S. Khalid, *Polymer* 48, 827–840, 2007. With permission.)

Layered silicates such as talc, mica, and smectic clays consist of regularly stacked layers, each having an average thickness of about 1 nm and a length of about 50 to 1000 nm. These types of layered silicates are classified as 2:1, as the individual layer is made up of three sheets—two on the outside and one in between. The outer sheets are made of interconnected SiO_4 tetrahedrals, whereas the inner one has an octahedral configuration. It contains metal cations, such as Al^{3+} and Mg^{2+}, and hydroxyl groups.

Various 2:1 layered silicates differ in the metal cation of the inner sheet. For example, talc contains Mg^{2+}, mica comprises Al^{3+}, and smectic clay has Fe^{2+} or Fe^{3+}. If the same metal as Mg^{2+} in talcum occupies all sites, the silicate is neutral and mechanical processes can easily separate the layers. In the silicate layers of mica or smectic clay Na^+, Ca^+, or K^+ act as countercharges in the gallery spacing, resulting in a strong ionic bond that is hard to break. This leads to a high-energy interlayer. Maine and Shepherd were the first to consider the possibility of using single clay layers as reinforcing fillers for polymer matrices [58]. However, they have expressed doubt about the feasibility of separating the clay clusters into their individual layers via simple mechanical action as in the conventionally used extruders.

Among the existing nanofillers, MMT clays are the most commonly employed for the production of nanocomposites. They are 2:1 layer silicates. The MMT particles are made up of layers having a length within 100 to 1000 nm, a thickness of 1 nm, and a high aspect ratio within 10 to 1000, thus possessing very high surface area for polymer/filler interaction.

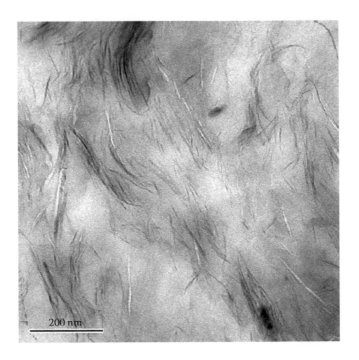

FIGURE 3.4

TEM photomicrograph of ethylene vinyl acetate nanocomposites containing 30 wt% of cupper-modified organo-clay cloisite C20A, prepared using a melt-compounding process at a temperature of 170°C, a screw rotation of 200 rpm, for a mixing time of 10 min. Modified organoclays were prepared by treating the organoclay suspensions in alcohol (ethanol or methanol) with a transition metal ion (cupper) solution in the same solvent. (From P. Nawani, P. Desai, M. Lundwall, M. Y. Gelfer, B. S. Hsiao, M. Rafailovich, A. Frenkel, A. H. Tsou, J. W. Gilman, and S. Khalid, *Polymer* 48, 827–840, 2007. With permission.)

The composites shown in Figures 3.5 and 3.6 having different clay contents were compounded using a twin-screw miniextruder designed by DSM-Research at a temperature of 230°C, a screw rotation speed of 80 rpm, under nitrogen atmosphere for a mixing time of 10 min. The polymers were previously dried overnight in a vacuum oven at 70°C to remove water. A two-step mixing procedure was used to prepare the polyvinylidene fluoride (PVDF) nanocomposites. Polymethyl methacrylate (PMMA)–organoclay master batches with different contents of silicate were first prepared by melt-mixing and were subsequently melt-mixed with PVDF. One-step mixing was also considered. (From N. Moussaif and G. Groeninckx, *Polymer* 44, 7899–7906, 2003. With permission.)

The interlayer distance may vary from 1 to 3 nm. MMT consists of gross agglomerates of 0.1 mm in diameter. Agglomerates are made up of primary particles of 1 to 10 um in size. A primary particle is composed of tachtoids (crystallites) of 0.1 to 1 um.

Silicates clay like hectonite, MMT, and saponite can be swollen in water. The ionic charges are thus weakened, and the gallery space is significantly increased. Because of the absence of intergallery counterion (cation), talc is hydrophobic and cannot be swollen in water. Korbee [38] used the water swelling of the clays in the presence of PA6 in the extruder to produce a PA6-clay nanocomposite.

Silicate clays as recovered from nature do not lead to the formation of nanocomposites by simply physically mixing them with a polymer. Except with few hydrophilic polymers like polyethylene oxide [49–52] or polyvinyl alcohol [53,54], which have good interaction with the inorganic, highly hydrophilic, layered silicates, the surfaces of the silicates have

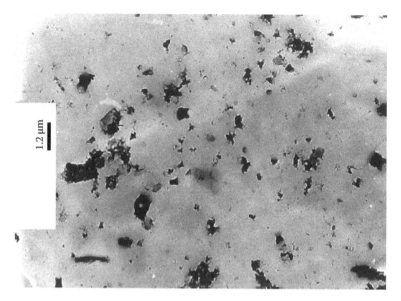

FIGURE 3.6
TEM photomicrograph of PMMA added with 5 wt% hydrophilic MMT. A micron-scale dispersion of primary particle aggregates is observed in the PMMA matrix, as a result of the high interfacial tension between the hydrophobic polymers and the hydrophilic montmorillonite clay. (From N. Moussaif and G. Groeninckx, *Polymer 44*, 7899–7906, 2003. With permission.)

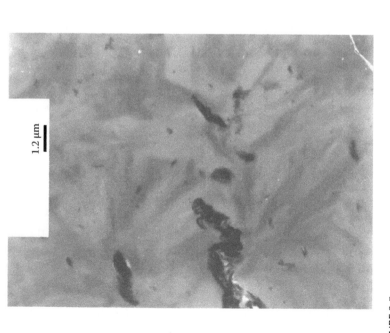

FIGURE 3.5
TEM photomicrograph of PVDF added with 5 wt% of hydrophilic MMT prepared by melt-extrusion. A micron-scale dispersion of primary particles aggregates is observed in the PVDF matrix, as a result of the high interfacial tension between the hydrophobic polymers and the hydrophilic montmorillonite clay. (From N. Moussaif and G. Groeninckx, *Polymer 44*, 7899–7906, 2003. With permission.)

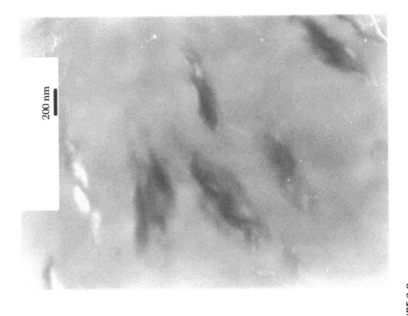

FIGURE 3.8
Higher-magnification TEM photomicrograph of PVDF added with 5 wt% of hydrophilic MMT prepared by melt-extrusion; same sample as in Figure 3.7. (From N. Moussaif and G. Groeninckx, *Polymer* 44, 7899–7906, 2003. With permission.)

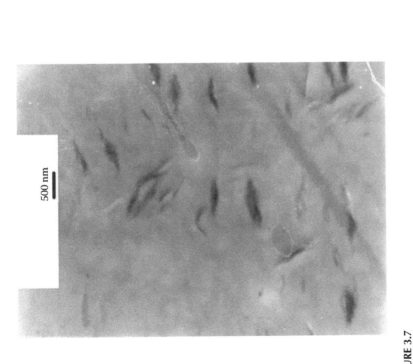

FIGURE 3.7
TEM photomicrograph of PVDF added with 5 wt% of hydrophilic MMT prepared by melt-extrusion. In sharp contrast to Figure 3.5, mixing PVDF with organophilic MMT leads to a relatively fine dispersion of silicate particles, but it failed to exfoliate the silicate layers. (From N. Moussaif and G. Groeninckx, *Polymer* 44, 7899–7906, 2003. With permission.)

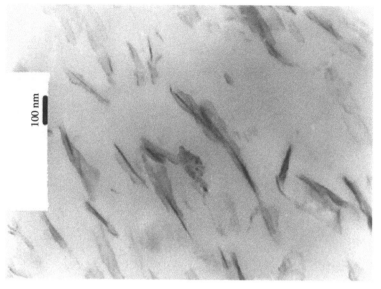

FIGURE 3.10
The higher-magnification image of PMMA nanocomposites shows individual layers as well as stacks containing parallel-oriented layers with various degrees of intercalation. The intercalation of PMMA chains between the silicate layers is enhanced by the strong polar interaction developed between the oxygen groups of the silicate and the oxygen groups of PMMA; same sample as in Figure 3.9. (From N. Moussaif and G. Groeninckx, *Polymer 44*, 7899–7906, 2003. With permission.)

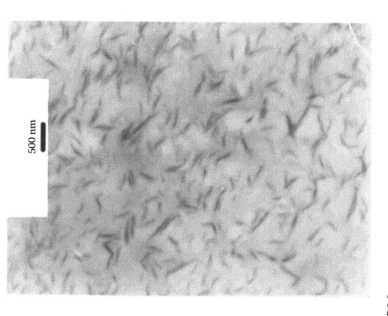

FIGURE 3.9
TEM photomicrograph of PVDF added with 5 wt% of hydrophilic MMT prepared by melt-extrusion. PMMA–organophilic MMT composites exhibit fine dispersion of nearly individual silicate platelets. (From N. Moussaif and G. Groeninckx, *Polymer 44*, 7899–7906, 2003. With permission.)

to be rendered organophilic in order to get intercalation or exfoliation when dispersed in most engineering polymers. This can be achieved by ion-exchange reactions with cationic surfactants including primary, secondary, tertiary, and quaternary alkylammonium or alkylphosphonium cations. This reduces the surface energy of the inorganic host and allows for an easy surface wetting by the polymer molecules. In addition, the alkylammonium or alkylphosphonium allows the attachment of reactive functional groups either for chemically tying the polymer molecules to the silicate layer or to provide sites for the initiation of *in situ* polymerization of appropriate monomers [55,56]. The organic cations should be able to replace the interlayer silicate cations, or the organic tails bearing functional groups should be capable of complexing them. When this modification is applied starting from an aqueous solution, the intercalation process leads to swollen clay interlayer distances within 1 to 4 nm. Water is removed without causing a decrease in the interlayer thickness. Polymer molecules can thus enter the interlayers, resulting in exfoliated clays and leading to the production of nanocomposites.

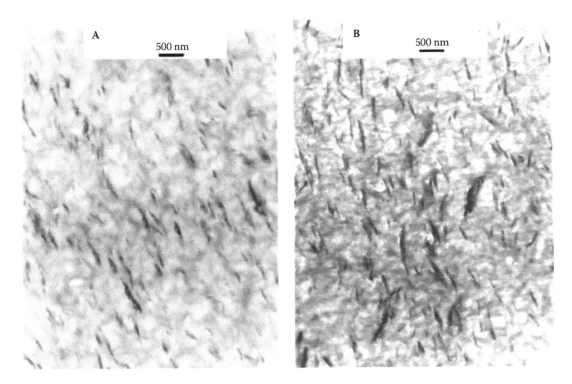

FIGURE 3.11

TEM photomicrographs of PVDF/PMMA–organophilic MMT nanocomposites containing 5 wt% of organophilic MMT. The content of PMMA in the miscible PVDF/PMMA blend matrix is as follows: (A): 5 wt%; (B): 10 wt%; (C): 15 wt%; (D): 20 wt%; (E): 40 wt%; and (F): 55 wt%. From (A), it is clear that blending PVDF with only 5 wt% PMMA leads to a regular and homogeneous dispersion of intercalated and delaminated silicate layers, compared with the neat PVDF–organophilic MMT composites (Figure 3.5). The premixing of PVDF with an increasing amount of PMMA from 10 to 55 wt% does not basically affect the nanomorphology dispersion (photomicrographs: B, C, D, E, and F). This observation emphasizes the beneficial effect of PMMA as a compatibilizer, which preferably intercalates between the silicate layers to the point where the silicate interlayer interactions become significantly weakened. Hence a partially exfoliated structure is generated as a result of the shear forces applied during melt-extrusion. (From N. Moussaif and G. Groeninckx, *Polymer* 44, 7899–7906, 2003. With permission.)

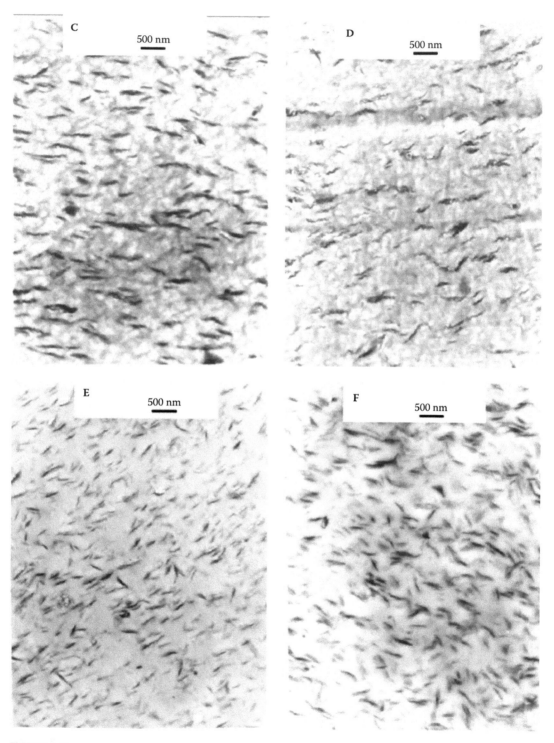

FIGURE 3.11
(*Continued*)

The question of how polymer molecules are able to diffuse inside the interlayer spacing of the clay, although their average radius of gyration is 10 to 20 times higher than the available 1 to 5 nm spacing of the interlayer gallery, has drawn the attention of numerous qualified research groups during the past decade.

In amorphous polymer-based nanocomposites this question is not very crucial because the polymer molecules can either be in the amorphous bulk matrix or be confined within the clay interlayers depending on the interactions between the clay surface and the matrix. In semicrystalline polymer matrices, however, the situation is much more complex, as three situations can be distinguished: the molecules (1) remain in the crystalline phase of the polymer matrix, (2) stay in the amorphous state, or (3) reside confined within the

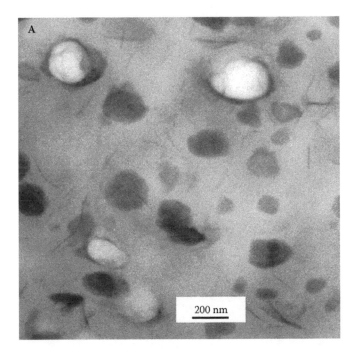

FIGURE 3.12
TEM images of nylon-6/unmodified clay/rubber nanocomposites. (A): nylon-6/VP-UFPRM (100/10 wt/wt); (B): nylon-6/A-UFPRM (100/10 wt/wt); and (C): nylon-6/S-UFPRM (100/10 wt/wt). These photomicrographs show the phase morphology of nylon-6/unmodified clay/rubber nanocomposites, which are prepared by melt-blending nylon-6 with a special compound powder, ultra-fine full-vulcanized powdered rubber/ montmorillonite (UFPRM). UFPRM consisting of ultra-fine full-vulcanized powdered rubber (UFPR) and unmodified clay is obtained after spray-drying the mixture of irradiated rubber latex and clay slurry. UFPR can be prepared by using irradiation-sensitive rubber latex as a raw material followed by irradiation cross-linking and drying before obtaining full-vulcanized powdered rubber with basically same-sized particles in rubber latex. The latex particles in most chosen rubber latexes are 50 to 150 nm in size, so we refer to all the full-vulcanized powdered rubber smaller than micron scale as ultrafine full-vulcanized powdered rubber. Based on clay slurry and butadiene styrene vinyl-pyridine, acrylate, and silicone rubber latex, three compound powders of UFPRM (UFPR/MMT = 4/1, wt/wt)—VP-UFPRM (butadiene styrene vinyl-pyridine UFPRM), A-UFPRM (acrylate UFPRM), and S-UFPRM (silicone UFPRM)—can be prepared. As shown in the TEM photomicrographs of (A): there are many black particles (i.e., UFPR particles) and a lot of dark lines (clay platelets) in the nylon-6/ VP-UFPRM nanocomposites. Most UFPR particles are well dispersed in the nylon-6 matrix and the silicate layers are well exfoliated in the nylon-6 matrix among the UFPR particles. Because acrylates and silicone rubber particles are hard for TEM to discern, only the exfoliated silicate layers can be seen in (B) and (C). (From W. Dong, X. Zhang, Y. Liu, Q. Wang, J. Gao, Z. Song, J. Lai, F. Huang, and J. Qiao, *Polymer* 42, 2515, 2005. With permission.)

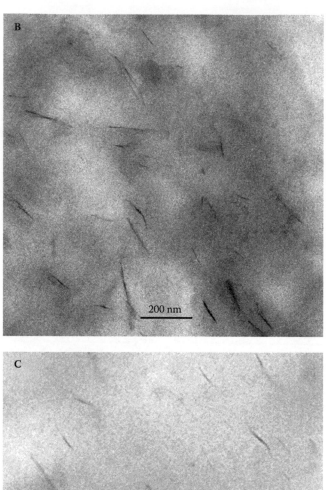

FIGURE 3.12
(*Continued*)

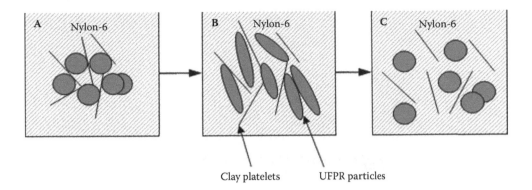

FIGURE 3.13

Schematic diagram depicting dispersion of UFPRM in nylon-6 during the melt-blending process. This figure is proposed to depict the dispersion mechanism of unmodified clay in nylon-6 during melt-blending. At the beginning of the melt-blending, the interpartitioned structure of UFPR and clay in UFPRM is retained in nylon-6 (A). Because the shear stress is easily transferred to the UFPR particles, the UFPR particles are elongated under the shear stress. At the same time, clay platelets are dispersed into the nylon-6 matrix with the elongation of UFPR particles (B). When UFPR particles are fixed into the melting nylon-6, the clay platelets are difficult to aggregate due to the partitioning effect of UFPR on clay platelets (C). It can be seen that the key point of silicate layers exfoliation is the special interpartitioned structure of UFPR and clay in UFPRM formed during spray-drying the mixture of clay slurry and rubber latex. Therefore, the UFPR can help unmodified clay to be exfoliated in nylon-6. (From W. Dong, X. Zhang, Y. Liu, Q. Wang, J. Gao, Z. Song, J. Lai, F. Huang, and J. Qiao, *Polymer* 42, 2515, 2005. With permission.)

interlayer spacing of the clay. Also an intermediate interphase can exist between the clay stacks and the crystalline lamellae of the polymer matrix.

The phenomenon of chain confinement is still a subject of scientific controversy. For example, an alkylammonium-modified flurohectorite silicate having an interlayer spacing of 2 nm has been dispersed in polystyrene [57]. The authors pretend that the polystyrene molecules diffuse into the interlayer gallery at a rate about 10 to 100 times higher than that in the bulk. Such a statement is hard to conceive knowing that the radius of gyration of polystyrene is equal to or higher than 100 nm. How is it possible to insert such large molecules, even in their fully extended state, into a 2 nm space where in fact the surfactant further restrains the entropy?

Furthermore, it is still not clearly elucidated what precise conditions lead to exfoliated layers and what causes only intercalation of chains into the gallery spacing of the stacked silicate layers. Many researchers consider the intercalation as due to a lack of interaction between the silicate layers and the polymer molecules of the matrix, whereas in most of the research reported so far, exfoliation is considered to result from high interaction between the polymer molecules and the organic surfactants used to modify the inorganic silicate layers. All the conclusions drawn are depicted from the measured dimensions of the interlayer spacing using X-ray diffraction (XRD) or in some instances transmission electron microscopy (TEM). Can we postulate that a simple intercalation is the intermediate nonequilibrium step of the process of exfoliation, or should it be accepted as a thermodynamic state of a process forcing a polymer molecule in between two layers of silicate without overcoming the energy barrier toward their long-range splitting into two individual layers (exfoliation)? An additional energy barrier necessary to overcome consists

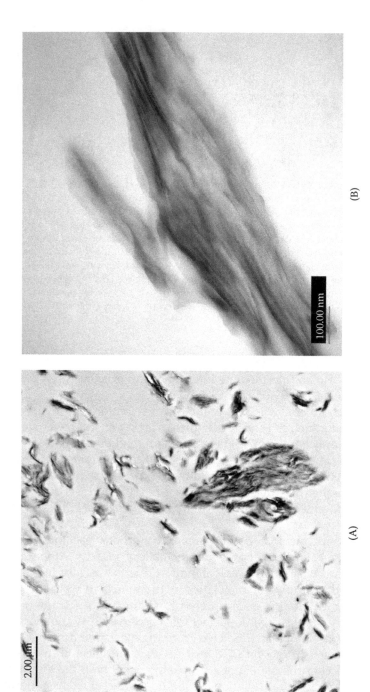

(A)

(B)

FIGURE 3.14

TEM photomicrographs of PMMA nanocomposites containing 5 wt% of the commercial organoclay cloisite 93A (methyl-dehydrogenated-tallow-ammonium modified montmorillonite; Southern Clay Products, Inc.). (A): low magnification; (B): high magnification, observation. The investigated samples were synthesized by free-radical polymerization of methyl methacrylate (MMA) (0.2 wt% tert-butyl peroxide as a thermal initiator, and 0.7 wt% 1-dodecanethiol as a chain transfer agent) containing organoclay, which was dispersed by 5 h of sonication in an ice bath, by gradual heating to 90°C with a 48-h hold. Sections (nominal thickness = 70 nm) were cut with a diamond knife and carbon coated. Imaging was carried out using a Philips CM12 TEM, 120 kV. (From K. R. Ratinac, R. G. Gilbert, L. Ye, A. S. Jones, and S. P. Ringer, *Polymer* 47, 6337–6361, 2006. With permission.)

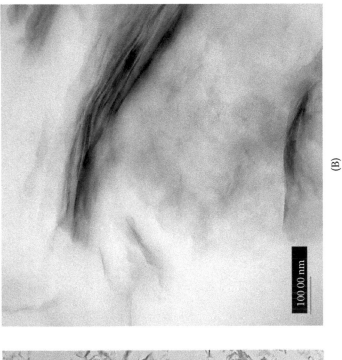

(B)

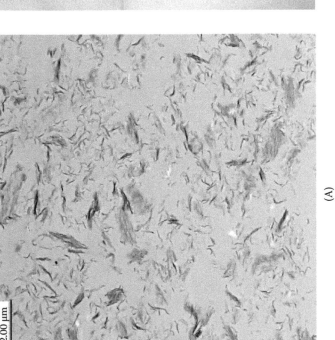

(A)

FIGURE 3.15

Transmission electron photomicrographs of a PMMA nanocomposite containing 5 wt% of the commercial organoclay cloisite 30B (bis-2-hydroxyethyl-methyl-tallow-ammonium modified montmorillonite; Southern Clay Products, Inc.). Sample synthesized by free-radical polymerization of MMA (0.2 wt% tert-butyl peroxide as a thermal initiator, and 0.7 wt% 1-dodecanethiol as a chain transfer agent) containing organoclay, which was dispersed by 5 h of sonication in an ice bath, by gradual heating to 90°C with a 48-h hold. Sections (nominal thickness = 70 nm) were cut with a diamond knife and carbon coated. Imaging: Philips CM12 TEM, 120 kV. (A): low magnification; (B): high magnification, observation. (From K. R. Ratinac, R. G. Gilbert, L. Ye, A. S. Jones, and S. P. Ringer, *Polymer 47*, 6337–6361, 2006. With permission.)

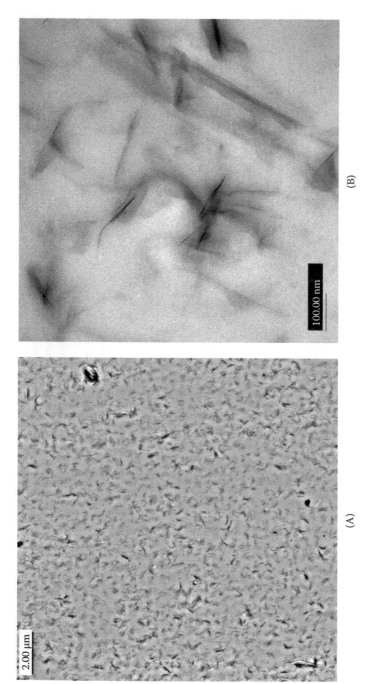

(A)

(B)

FIGURE 3.16

Transmission electron photomicrographs of a PMMA (Plexiglas 7H, $M_w \sim 150$ kD; Röhm GmbH & Co. KG, Darmstadt, Germany) nanocomposite containing 5 wt% of the commercial organoclay cloisite 93A (methyl-dehydrogenated-tallow-ammonium modified montmorillonite; Southern Clay Products, Inc.). (A): low magnification; (B): high magnification, observation. Sample synthesized by melt-compounding of oven-dried PMMA pellets with the organoclay at approximately 200°C in a ZSK-30 intermeshing, corotating, twin-screw extruder (L/D = 30, L = 880 mm, speed = 250 rpm; Werner and Pfleiderer, Ramsey, New Jersey). Samples were extruded through a 3-mm circular die and water-cooled. Transverse sections (nominal thickness = 70 nm) were cut with a diamond knife and carbon coated. Imaging: Philips CM12 TEM, 120 kV. (From K. R. Ratinac, R. G. Gilbert, L. Ye, A. S. Jones, and S. P. Ringer, *Polymer* 47, 6337–6361, 2006. With permission.)

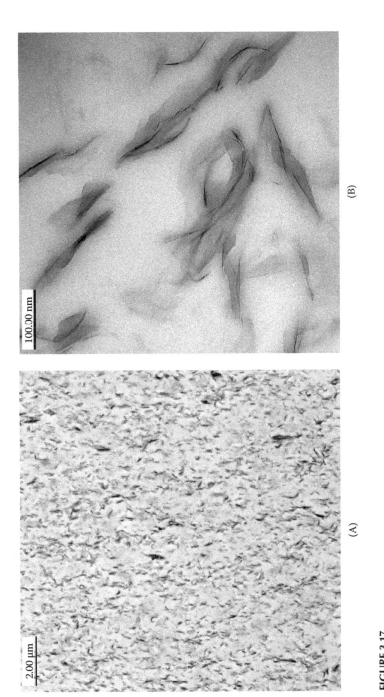

FIGURE 3.17

Transmission electron photomicrographs of PMMA (Plexiglas 7H, $M_w \sim 150$ k; Röhm GmbH & Co. KG, Darmstadt, Germany) nanocomposite containing 5 wt% of the commercial organoclay cloisite 30B (bis-2-hydroxyethyl-methyl-tallow-ammonium modified montmorillonite; Southern Clay Products, Inc.). (A): low magnification; (B): high magnification, observation. Sample synthesized by melt-compounding of oven-dried PMMA pellets with the organoclay at approximately 200°C in a ZSK-30 intermeshing, corotating, twin-screw extruder (L/D = 30, L = 880 mm, speed = 250 rpm; Werner and Pfleiderer, Ramsey, New Jersey). Samples were extruded through a 3-mm circular die and water-cooled. Transverse sections (nominal thickness = 70 nm) were cut with a diamond knife and carbon coated. Imaging: Philips CM12 TEM, 120 kV. (From K. R. Ratinac, R. G. Gilbert, L. Ye, A. S. Jones, and S. P. Ringer, *Polymer* 47, 6337–6361, 2006. With permission.)

FIGURE 3.19
TEM photomicrograph of a PMMA nanocomposite containing 5 wt% of size-fractionated bentonite clay (montmorillonite) modified with the polymerizable quaternary surfactant [2-(methacryloyloxy)ethyl]tetradecyldimethylammonium bromide (at 1.6 CEC). This figure and Figure 3.18 illustrate the relative differences in morphology in poly(methyl methacrylate)-clay samples caused by the use of a polymerizable modifier. (From K. R. Ratinac, R. G. Gilbert, L. Ye, A. S. Jones, and S. P. Ringer, *Polymer* 47, 6337–6361, 2006. With permission.)

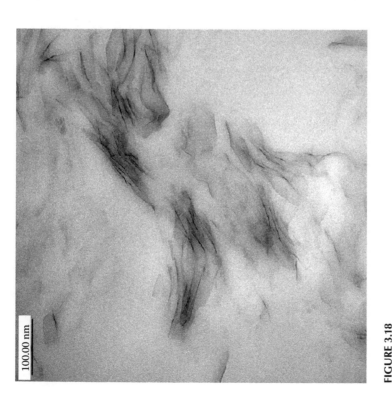

FIGURE 3.18
TEM photomicrograph of a PMMA nanocomposite containing 5 wt% of size-fractionated bentonite clay (montmorillonite) modified with cetyl trimethyl ammonium bromide (at 1.9 cation exchange capacity [CEC]). (From K. R. Ratinac, R. G. Gilbert, L. Ye, A. S. Jones, and S. P. Ringer, *Polymer* 47, 6337–6361, 2006. With permission.)

FIGURE 3.20

TEM photomicrograph of the "nanocomposite" formed via a drop-wise addition of an aqueous dispersion of montmorillonite platelets into an aqueous suspension of cationic PMMA latex particles under high-speed mechanical mixing. Latex synthesized by emulsion polymerization of MMA and 0.5 wt% (relative to MMA) of [2-(methacryloyloxy)ethyl]trimethylammonium chloride with AIBA as a thermal initiator (as per X. Y. Huang and W. J. Brittain, *Macromolecules* 34(10), 3255–3260, 2001). The combined suspension was diluted and placed on a Formvar-coated TEM grid and dried. Imaging: Philips CM12 TEM, 120 kV. (From K. R. Ratinac, R. G. Gilbert, L. Ye, A. S. Jones, and S. P. Ringer, *Polymer* 47, 6337–6361, 2006. With permission.)

of disentangling the entangled polymer molecules before they get inserted in the interspacing of the silicate layers. One can imagine all the possible figures. Has any investigation been devoted to the use of oligomers instead of high-molecular-weight (high-MW) polymers to check whether the low-MW species can result only in an intercalated state, whereas the high-MW chains cause the destruction of organized silicate layers, or are they (without the existence of low species) unable to cause even intercalation? The idea behind this statement is that in a polydisperse polymer, the low-MW chains can diffuse merely into the gallery of the silicates, facilitating the progressive insertion of larger and larger molecules. If this hypothesis is verified, then one would expect that with the same type of clay broader MWD polymers would be more efficient than narrow MWD polymers in generating an exfoliated state. Note that in open air–substrate interfaces, a low-MW polymer diffuses faster to the surface.

The present chapter is devoted mainly to the phase morphology reported for silicate clay–based nanocomposites. Many of the selected TEM photomicrographs show exfoliated and intercalated states of clay in various polymer matrices.

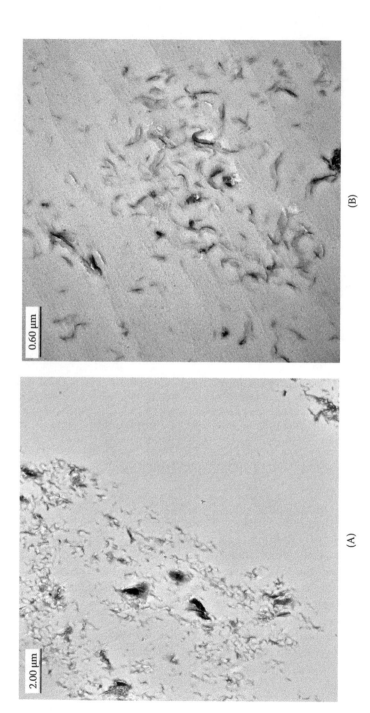

(A)

(B)

FIGURE 3.21

TEM photomicrographs of PMMA-clay nanocomposites formed by hot-pressing, at 180°C, the dried latex-clay "nanocomposite" shown in Figure 3.20 (as per X. Y. Huang and W. J. Brittain, *Macromolecules* 34(10), 3255–3260, 2001). The nanocomposite contains 5 wt% of size-fractionated bentonite clay (montmorillonite). (A): low magnification; (B): high magnification. These figures illustrate the relative differences in morphology in poly(methyl methacrylate)-clay nanocomposites formed from unmodified clay and cationic PMMA latex made by emulsion polymerization (as per the method of X. Y. Huang and W. J. Brittain, *Macromolecules* 34(10), 3255–3260, 2001). (From K. R. Ratinac, R. G. Gilbert, L. Ye, A. S. Jones, and S. P. Ringer, *Polymer* 47, 6337–6361, 2006. With permission.)

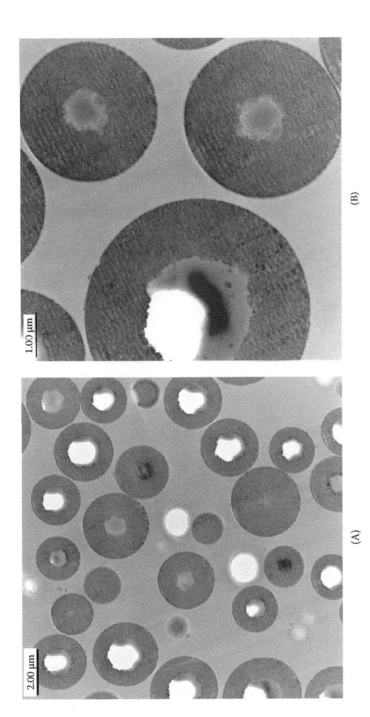

(A)

(B)

FIGURE 3.22

TEM photomicrographs of core-shell latex particles made by impregnation of hollow poly(divinylbenzene) shells by molten N-isopropyl acrylamide (NIPAM) and subsequent gamma irradiation to polymerize the NIPAM. The particles were embedded in Spur's resin, were ultramicrotomed to a nominal section thickness of 70 nm by diamond knife, and then were stained with RuO$_4$ vapor. Imaging has been performed using a Philips CM12 TEM, 120 kV. (A): low magnification; (B): high magnification. (From T. Brand, K. Ratinac, J. V. Castro, and R. G. Gilbert, *J. Polym. Sci. A: Polym. Chem.* 42, 5706–5713, 2004. With permission.)

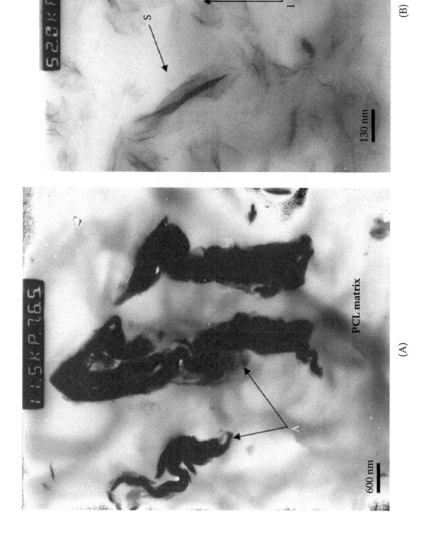

FIGURE 3.23

TEM photomicrographs of polycaprolactone (PCL)/cloisite Na$^+$(95/5) (A); and PCL/cloisite 30B(95/5) (B) composites. I: individual layer; S: stack of layers; A: aggregate of clays. (From L. Wang, *Localization of Silicates Clay in Poly-α-caprolactone (PCL), Polyethylene Oxide (PEO) Immiscible Polymer Blends*, master's thesis, Katholieke Universiteit Leuven, Belgium, 2003, under the supervision of G. Groeninckx, C. Harrats, and N. Moussaif.)

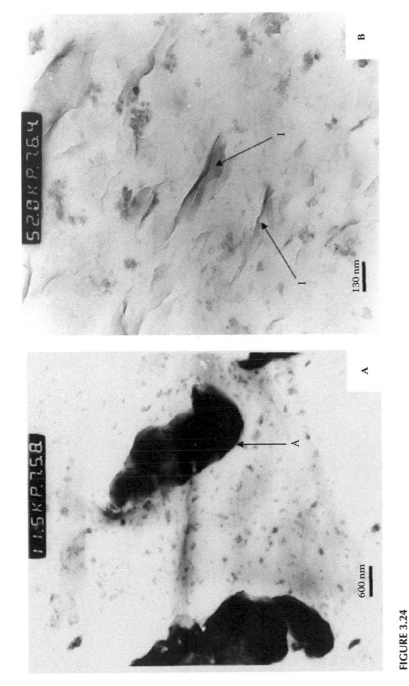

FIGURE 3.24

TEM photomicrographs of polyethylene oxide (PEO)/cloisite Na+(95/5) (A); and (B): PEO/cloisite 30B(95/5) composites. I: individual layer; A: aggregate of clays. (From L. Wang, *Localization of Silicates Clay in Poly-α-caprolactone (PCL), Polyethylene Oxide (PEO) Immiscible Polymer Blends*, master's thesis, Katholieke Universiteit Leuven, Belgium, 2003, under the supervision of G. Groeninckx, C. Harrats, and N. Moussaif.)

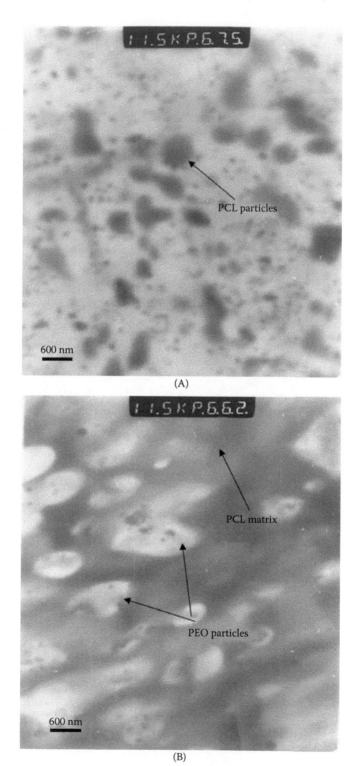

FIGURE 3.25
TEM photomicrographs of 80/20 wt/wt (A), and 20/80 wt/wt (B), melt-mixed PCL/PEO blends. (From Z. Fang, C. Harrats, N. Moussaif, and G. Groeninckx, *J. Appl. Polym. Sci.* 105, 3125, 2007. With permission.)

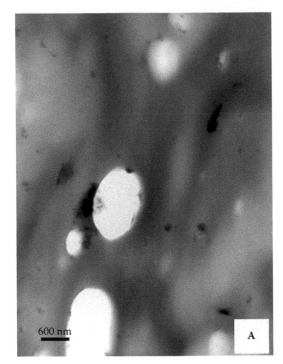

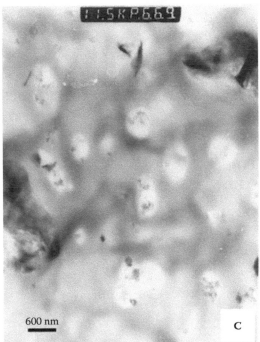

FIGURE 3.26

TEM photomicrographs of 80/20 wt/wt melt-mixed PCL/PEO blends with 1 wt% (A), 2 wt% (B), and 5 wt% (C), of cloisite Na[+]. (From L. Wang, *Localization of Silicates Clay in Poly-α-caprolactone (PCL), Polyethylene Oxide (PEO) Immiscible Polymer Blends*, master's thesis, Katholieke Universiteit Leuven, Belgium, 2003, under the supervision of G. Groeninckx, C. Harrats, and N. Moussaif.)

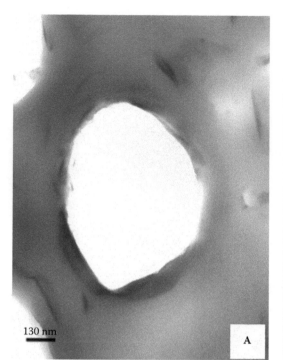

FIGURE 3.27
TEM photomicrographs of 80/20 PCL/PEO blends with 1 wt% (A), 2 wt% (B), and 5 wt% (C), of cloisite 30B. (From L. Wang, *Localization of Silicates Clay in Poly-ε-caprolactone (PCL), Polyethylene Oxide (PEO) Immiscible Polymer Blends*, master's thesis, Katholieke Universiteit Leuven, Belgium, 2003, under the supervision of G. Groeninckx, C. Harrats, and N. Moussaif.)

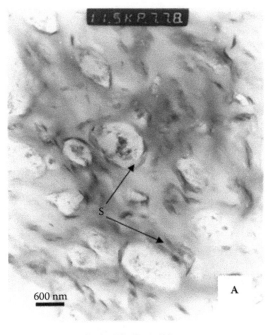

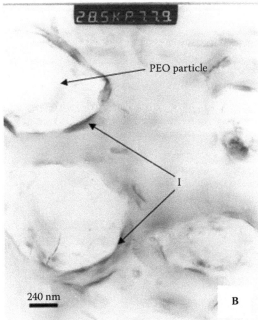

FIGURE 3.28

TEM photomicrographs of the 80/20 PCL/PEO blends with 5 wt% of cloisite 30B, prepared by two-step melt-extrusion. PCL was premixed with cloisite 30B first and then mixed with PEO. (A): low magnification; (B): higher magnification. (From L. Wang, *Localization of Silicates Clay in Poly-ε-caprolactone (PCL), Polyethylene Oxide (PEO) Immiscible Polymer Blends*, master's thesis, Katholieke Universiteit Leuven, Belgium, 2003, under the supervision of G. Groeninckx, C. Harrats, and N. Moussaif.)

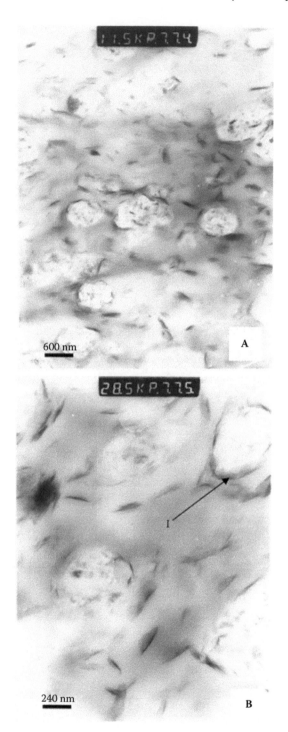

FIGURE 3.29
TEM photomicrographs of the 80/20 PCL/PEO blends with 5 wt% of cloisite 30B, prepared by a two-step melt-extrusion. PEO was premixed with cloisite 30B first and then mixed with PCL. (A): low magnification; (B): higher magnification. (From L. Wang, *Localization of Silicates Clay in Poly-ε-caprolactone (PCL), Polyethylene Oxide (PEO) Immiscible Polymer Blends,* master's thesis, Katholieke Universiteit Leuven, Belgium, 2003, under the supervision of G. Groeninckx, C. Harrats, and N. Moussaif.)

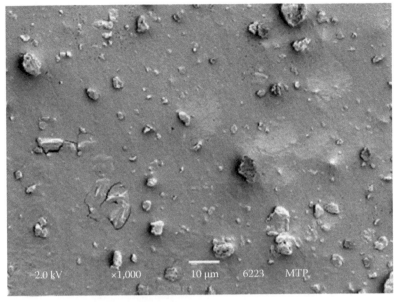

(A)

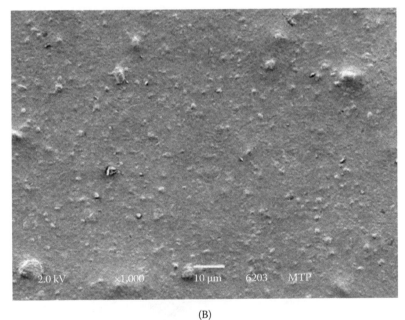

(B)

FIGURE 3.30
SEM photomicrographs taken from the etched surface of PP nanocomposites containing 2 vol.% organophilic clay: (A) polypropylene (PP)/organophilized montmorillonite clay (OMMT), (B) PP/OMMT/maleic anhydride modified polypropylene (MAPP) (20 vol.%). (From L. Szazdi, A. Abranyi, J. G. Vancso, and B. Pukanszky, *Macromol. Mater. Eng.* 291, 856–868, 2006. With permission.)

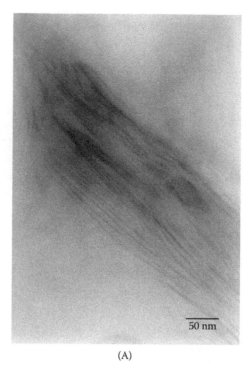

(A)

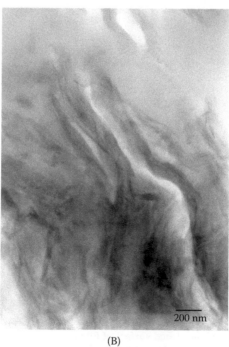

(B)

FIGURE 3.31
TEM photomicrographs taken from PP/OMMT/MAPP nanocomposites. (A): 3 vol.% silicate, 20 vol.% MAPP; (B): 2 vol.% silicate, 30 vol.% MAPP. (From L. Szazdi, A. Abranyi, J. G. Vancso, and B. Pukanszky, *Macromol. Mater. Eng.* 291, 856–868, 2006. With permission.)

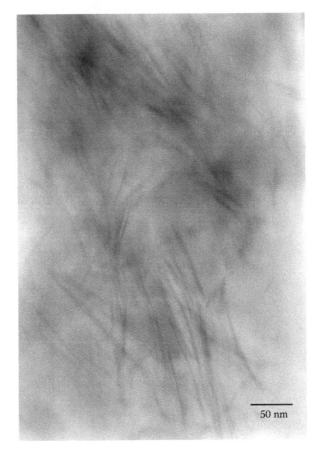

FIGURE 3.32
Interaction of silicate layers at small clay content of as low as 0.5 vol.% and MAPP content as large as 60 vol.%. The photomicrograph reveals possible network formation of silicate sheets. (From L. Szazdi, A. Abranyi, J. G. Vancso, and B. Pukanszky, *Macromol. Mater. Eng.* 291, 856–868, 2006. With permission.)

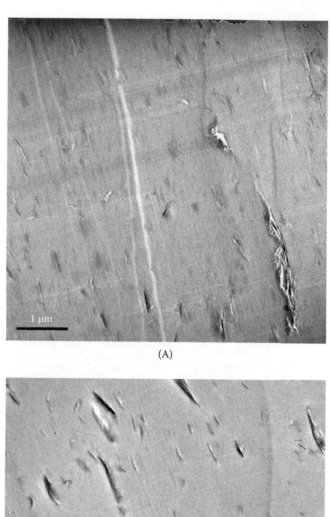

(A)

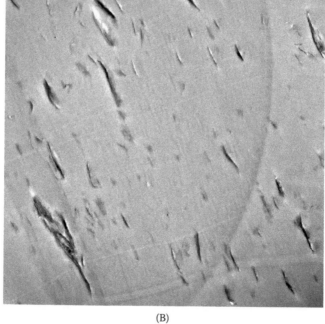

(B)

FIGURE 3.33
TEM photomicrographs viewed along the transverse direction (TD) at low magnification, 3000×, for the nanocomposites based on (A) PS, (B) SAN-2, (C) SAN-13.5, (D) SAN-25, (E) SAN-38, and (F) SAN-58. (From H. A. Stretz and D. R. Paul, *Polymer* 47, 8123, 2006. With permission.)

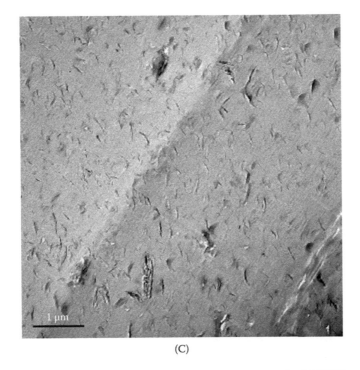

(C)

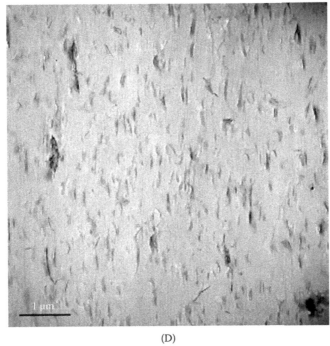

(D)

FIGURE 3.33
(*Continued*)

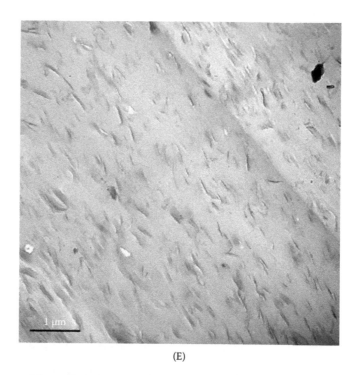

(E)

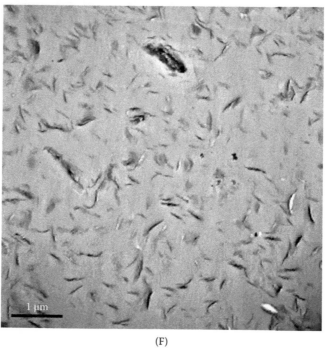

(F)

FIGURE 3.33
(*Continued*)

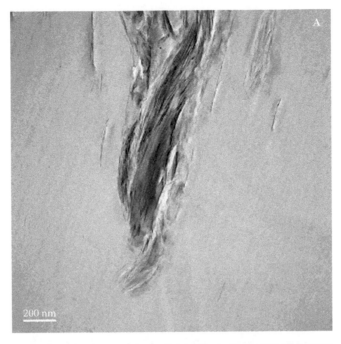

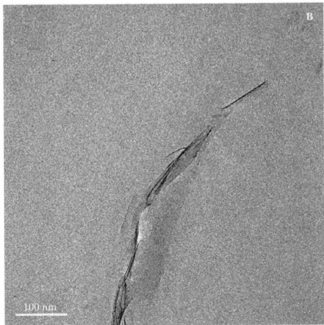

FIGURE 3.34
TEM photomicrographs of the polystyrene (PS)/$M_3(C_{18})_1$ composite, 3.2% MMT, viewed along the transverse direction (TD) (A, B), and the flow direction (FD) (C, D). (From H. A. Stretz and D. R. Paul, *Polymer* 47, 8123, 2006. With permission.)

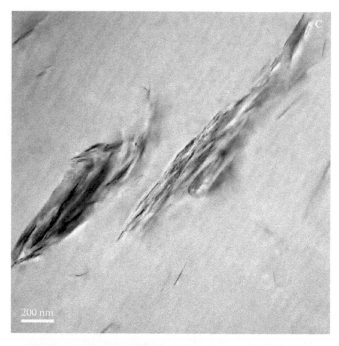

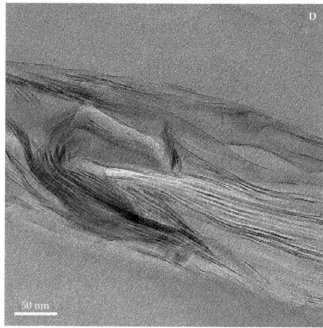

FIGURE 3.34
(*Continued*)

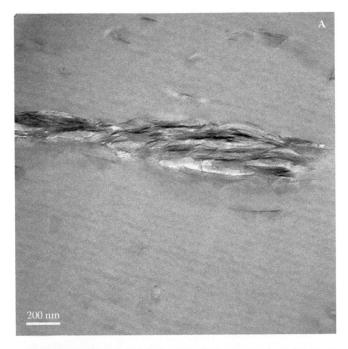

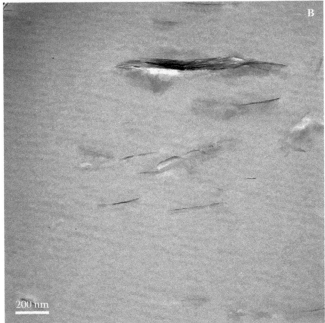

FIGURE 3.35

TEM photomicrographs of the SAN-2/$M_3(C_{18})_1$ composite, 3.2% MMT, viewed along the transverse direction (TD). (A, B, C: same magnification but different locations of observation; D: higher magnification). (From H. A. Stretz and D. R. Paul, *Polymer* 47, 8123, 2006. With permission.)

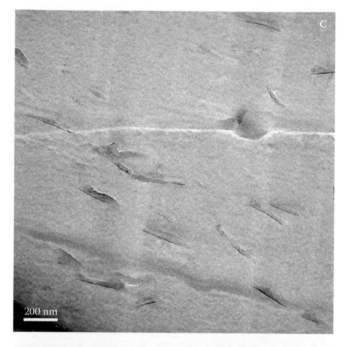

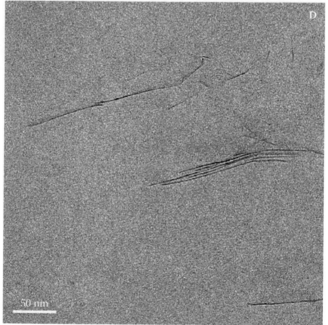

FIGURE 3.35
(*Continued*)

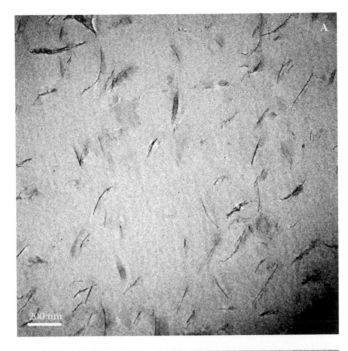

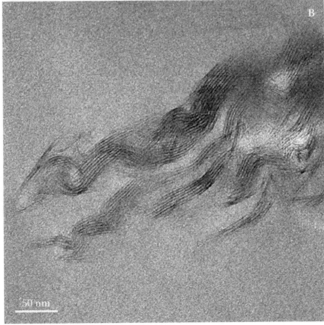

FIGURE 3.36
TEM photomicrographs of the SAN-13.5/$M_3(C_{18})_1$ composite, 3.2% MMT, viewed along the transverse direction (TD) (A, B), and the flow direction (FD) (C, D). (From H. A. Stretz and D. R. Paul, *Polymer* 47, 8123, 2006. With permission.)

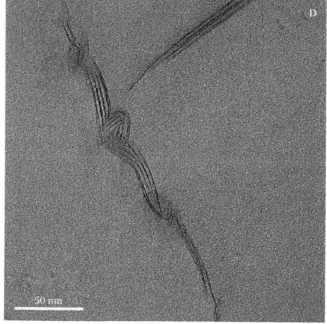

FIGURE 3.36
(*Continued*)

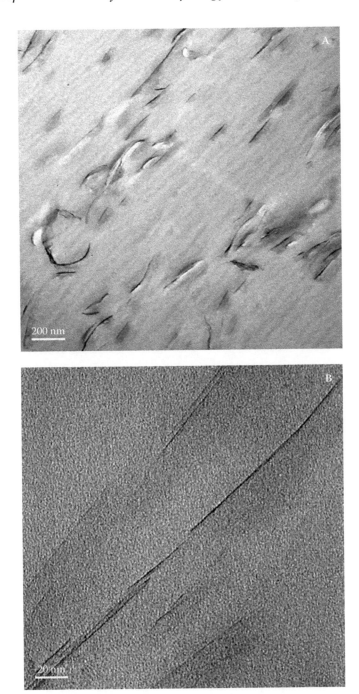

FIGURE 3.37
TEM photomicrographs of the SAN-25/M$_3$(C$_{18}$)$_1$ composite, 3.2% MMT, viewed along the transverse direction (TD) (A, B), and the flow direction (FD) (C, D). (From H. A. Stretz and D. R. Paul, *Polymer* 47, 8123, 2006. With permission.)

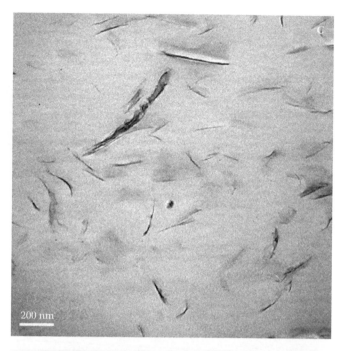

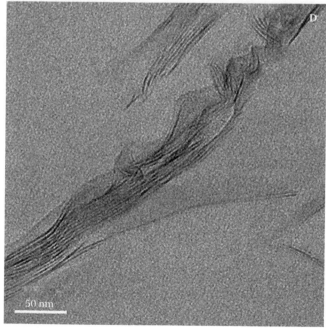

FIGURE 3.37
(*Continued*)

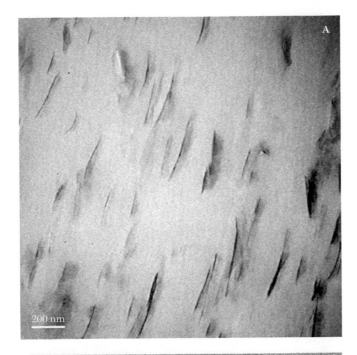

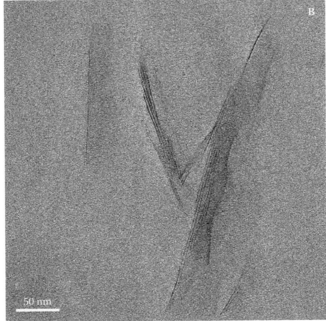

FIGURE 3.38

TEM photomicrographs of the SAN-38/M$_3$(C$_{18}$)$_1$ composite, 3.2% MMT, viewed along the transverse direction (TD) (A, B), and the flow direction (FD) (C, D). (From H. A. Stretz and D. R. Paul, *Polymer* 47, 8123, 2006. With permission.)

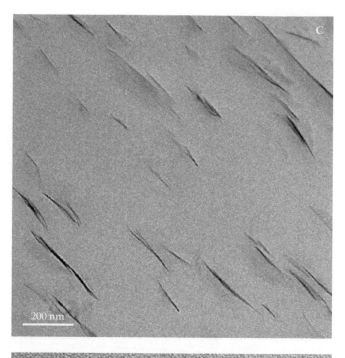

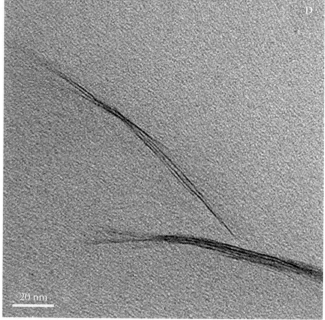

FIGURE 3.38
(*Continued*)

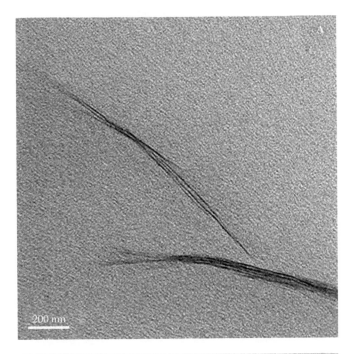

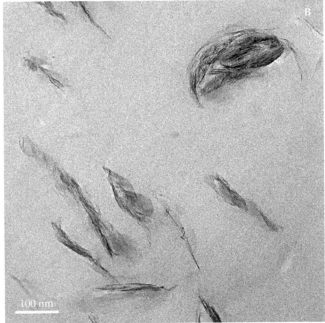

FIGURE 3.39
TEM photomicrographs of the SAN-58/M$_3$(C$_{18}$)$_1$ composite, 3.2% MMT, viewed along the transverse direction (TD) (A, B), and the flow direction (FD) (C, D). (From H. A. Stretz and D. R. Paul, *Polymer* 47, 8123, 2006. With permission.)

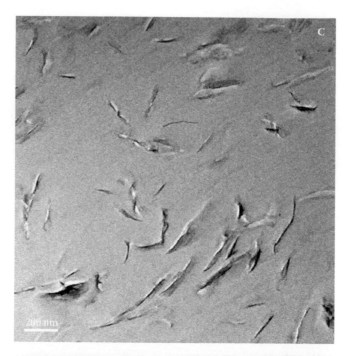

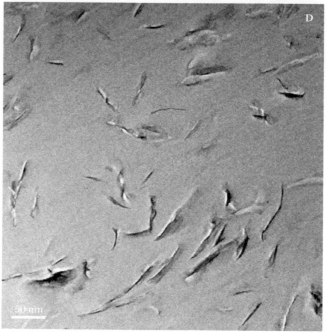

FIGURE 3.39
(*Continued*)

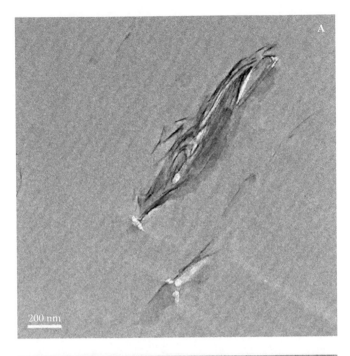

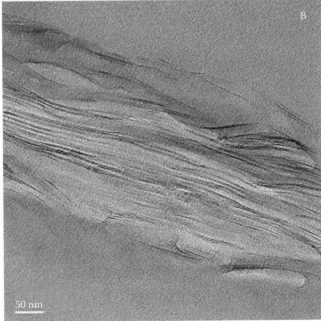

FIGURE 3.40
TEM photomicrographs of the PS/M$_3$(C$_{18}$) composite, 3.2% MMT, viewed along the flow direction. (A): low magnification; (B): high magnification. (From H. A. Stretz and D. R. Paul, *Polymer* 47, 8527, 2006. With permission.)

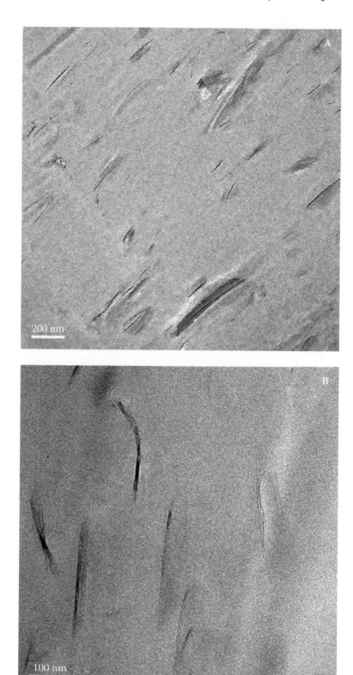

FIGURE 3.41

TEM photomicrographs of the SMA-14/$M_3(C_{18})$ composite (A, B) versus the SAN-13.5/$M_3(C_{18})$ composite, 3.2% MMT (C, D), viewed along the flow direction. (From H. A. Stretz and D. R. Paul, *Polymer* 47, 8527, 2006. With permission.)

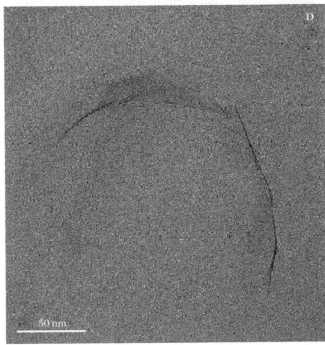

FIGURE 3.41
(*Continued*)

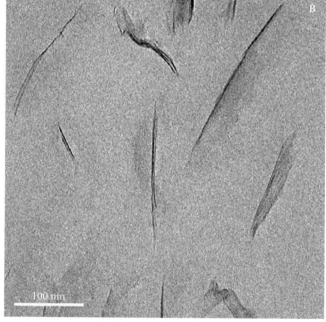

FIGURE 3.42
TEM photomicrographs of the SMA-25/$M_3(C_{18})$ composite (A, B) versus the SAN-25/$M_3(C_{18})$ composite (C, D), 3.2% MMT, viewed along the flow direction. (From H. A. Stretz and D. R. Paul, *Polymer* 47, 8527, 2006. With permission.)

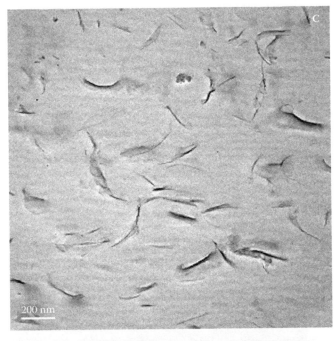

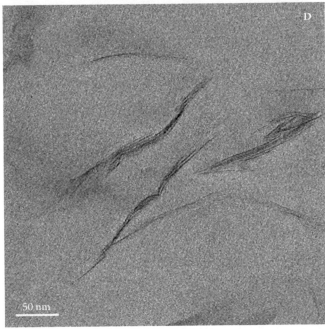

FIGURE 3.42
(*Continued*)

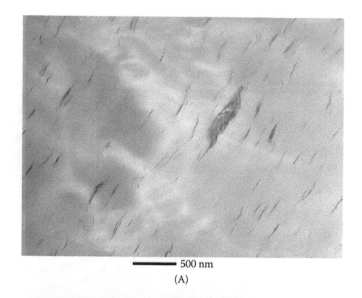

500 nm

(A)

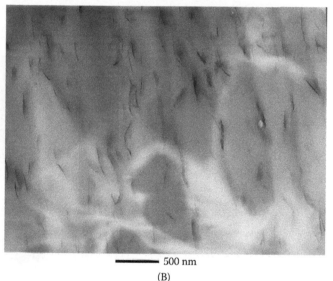

500 nm

(B)

FIGURE 3.43
TEM photomicrographs of the nanocomposites: 0/1–2% 15A (A); 7/3–2% 15A (B); 1/0–2% 15A (C); the same magnification for the three samples. The clays are dispersed quite finely and homogeneously in the matrix, in the form of single, double, triple, and multiple layer stacks; hence, the organoclay has been partly exfoliated. However, the quality of dispersion varies with the low-crystallinity polypropylene (LCPP) content. The nanocomposite with LCPP has the finest distribution with a greater number of smaller stacks (A). The NC with high-crystallinity polypropylene (HCPP) has the poorest dispersion with the greatest number of larger stacks (C). The nanocomposite based on the blend of HCPP and LCPP shows a quality of dispersion falling between the others (B), but the greater the amount of LCPP, the better the dispersion. (From M.-T. Ton-That, W. Leelapornpisit, L. A. Utracki, F. Perrin-Sarazin, J. Denault, K. C. Cole, and M. N. Bureau, *Polym. Eng. Sci.* 46, 1085, 2006. With permission.)

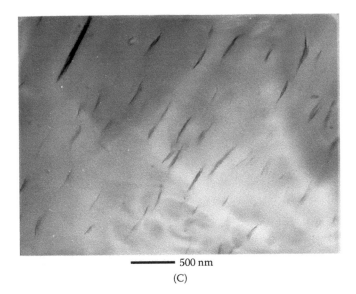

500 nm

(C)

FIGURE 3.43
(*Continued*)

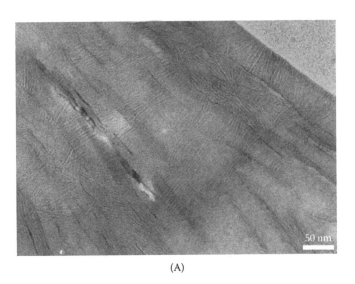

50 nm

(A)

FIGURE 3.44
TEM photomicrographs of PA6/clay nanocomposites: (A): +4 wt% clay; (B): +8 wt% clay; and (C): +12 wt% clay. These images reveal that depending on the clay loading (4, 8, and 12 wt%, which leads to an average clay platelet spacing of 61, 39, and 28 nm, respectively), the crystalline lamellae orientation of the PA6 is significantly different. Note that this interlayer spacing seems particularly much larger than what is commonly reported in nanocomposite reports in literature. For the low clay content (4 wt%), the PA6 crystalline lamellae were found to be mainly perpendicular to the clay layers. However, for the 12 wt% clay, having an interplatelet spacing of 28 nm, almost all the crystalline lamellae were parallel to the clay longitudinal axis. For the intermediate clay loading of 8 wt%, both perpendicular and parallel orientations of the PA6 crystalline lamellae were observed. Above 50 nm clay platelet spacing, the PA6 lamellae are mainly oriented perpendicular to the clay layers, whereas below a clay platelet spacing of 30 nm, the nylon-6 lamellae are oriented parallel to the clay layer axis. (From L. Yongjin and S. Hiroshi, *J. Polym. Sci., Part B: Polym. Phys.* 44, 284, 2006. With permission.)

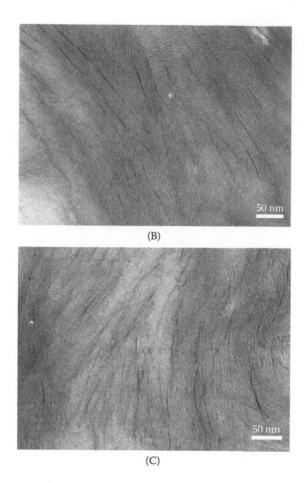

(B)

(C)

FIGURE 3.44
(*Continued*)

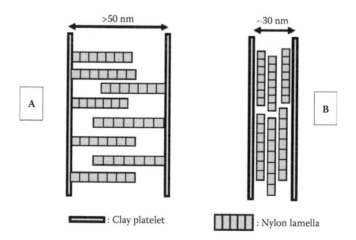

: Clay platelet   : Nylon lamella

FIGURE 3.45
Schematic representation of the PA6 lamellar morphologies between the clay platelets: (A) the spacing is larger than 50 nm, and (B) the spacing is less than 30 nm. (From L. Yongjin and S. Hiroshi, *J. Polym. Sci., Part B: Polym. Phys.* 44, 284, 2006. With permission.)

FIGURE 3.46

TEM photomicrographs for EVA9 (having 9% VA)/C15A clay nanocomposites: EVA9 +2.5 wt% C15A (A), EVA9 +5 wt% C15A (B), and EVA9 +7.5 wt% (C). The higher electron density of the silicates relative to the EVA matrix gives them a much darker appearance. The images show the presence of tactoids that are approximately 200 nm thick. It can be seen that increasing the clay concentration increases the thickness of the tactoids, suggesting that for the EVA9-cloisite 15A system, the clay platelets are not dispersed by the low matrix polarity. However, the tactoids themselves gain some degree of disorder, which is strongly affected by the clay concentration. (From D. S. Chaudhary, R. Prasad, R. K. Gupta, and S. N. Bhattacharya, *Thermochim. Acta* 433, 187–195, 2005. With permission.)

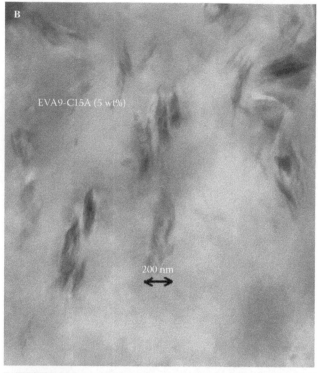

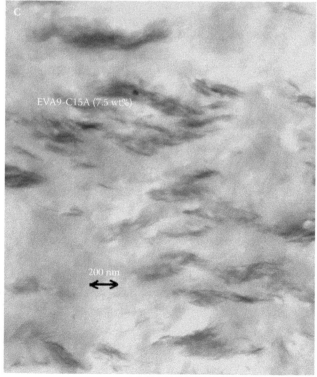

FIGURE 3.46
(*Continued*)

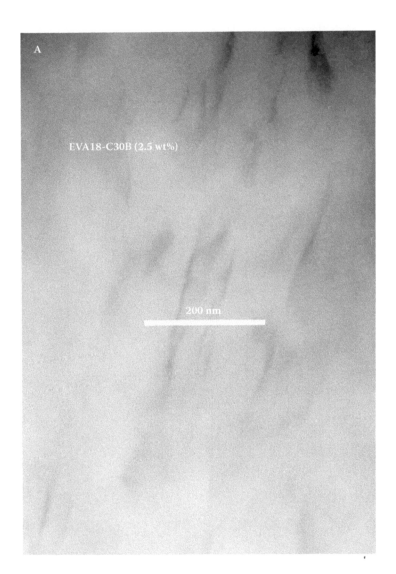

FIGURE 3.47

TEM photomicrographs for EVA18 (containing 18% VA)/clay (C30B grade) nanocomposites: EVA18 +2.5 wt% C30B (A), EVA18 +5 wt% C30B (B), and EVA18 +7.5 wt% C30B (C). Mixed intercalated/exfoliated morphologies are developed. The presence of stacks of silicate layers at higher concentration shows the presence of inter-calation, where few platelets are grouped together but possess random orientations, whereas at lower clay concentrations (2.5 and 5%), the clay platelets are scattered individually while some tactoid-structural order is preserved, suggesting mixed exfoliation. Larger particles could also be observed, and these may be tactoids that have not been dispersed well enough. (From D. S. Chaudhary, R. Prasad, R. K. Gupta, and S. N. Bhattacharya, *Thermochim. Acta* 433, 187–195, 2005. With permission.)

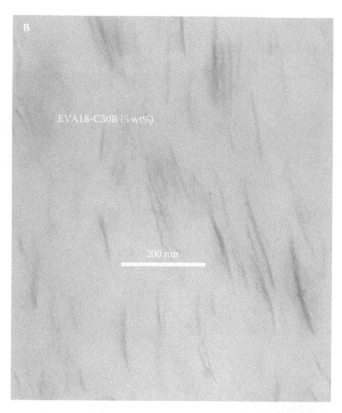

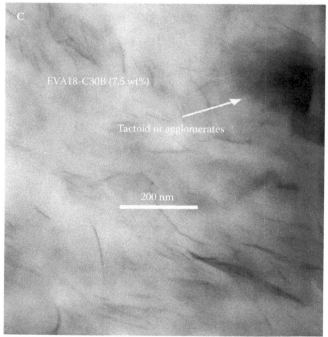

FIGURE 3.47
(*Continued*)

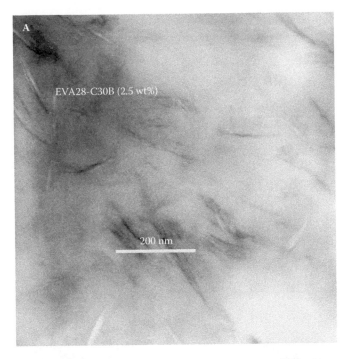

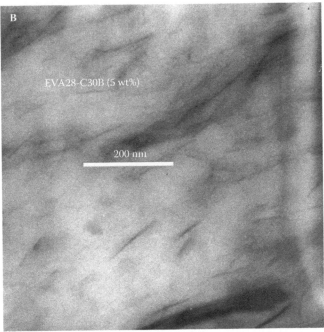

FIGURE 3.48

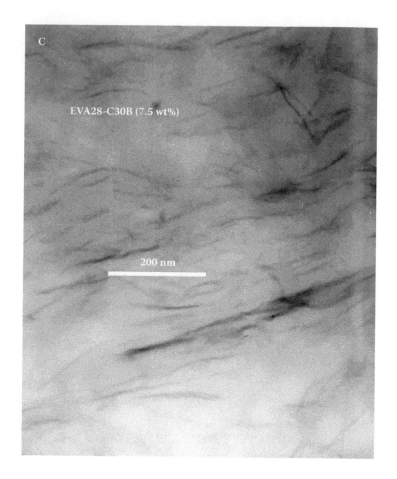

FIGURE 3.48 (*Continued*)
TEM photomicrographs for EVA28 (containing 28% VA)/clay (C30B grade) nanocomposites: EVA28 +2.5 wt% C30B (A), EVA28 +5 wt% C30B (B), and EVA28 +7.5 wt% C30B (C). These images reveal similar results as those obtained for EVA18 nanocomposites (Figure 3.47) in that they exhibit mixed intercalated/exfoliated morphologies. From these images, exfoliated individual layers could be seen interspersed with silicate stacks that are a few layers thick. The TEM images also show that with increased polarity of the matrix (28% VA), there is greater clay–polymer interaction, which is not significantly affected by increasing silicate loading from 2.5 wt% to 7.5 wt%, as opposed to EVA9 (Figure 3.46) or EVA18 (Figure 3.47). (From D. S. Chaudhary, R. Prasad, R. K. Gupta, and S. N. Bhattacharya, *Thermochim. Acta* 433, 187–195, 2005. With permission.)

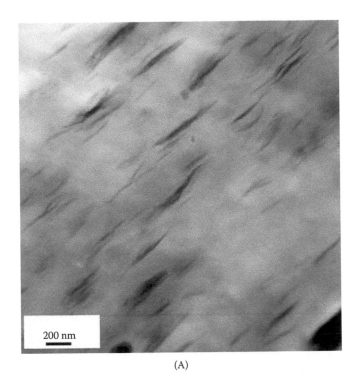

(A)

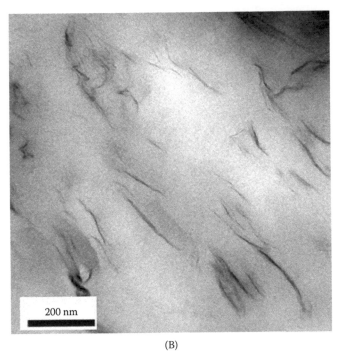

(B)

FIGURE 3.49

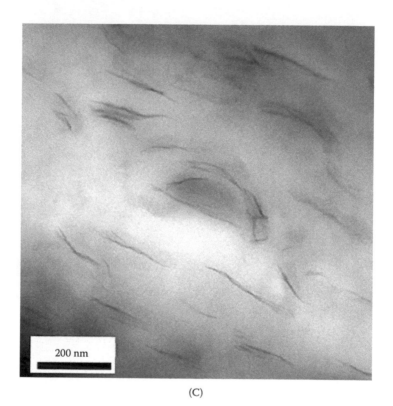

(C)

FIGURE 3.49 (*Continued*)
TEM photomicrographs of nanocomposites prepared from organoclay containing two long alkyl tails, M2(HT)2 dimethyl bis(hydrogenated tallow) and LDPE (A), ethylene/methacrylic acid copolymer containing 3.9 wt% methacrylic acid (EMAA-1) (B), and ethylene/methacrylic acid copolymer containing 8.9 wt% methacrylic acid (EMAA-2) (C). The concentration of MMT in all cases is 2.5 wt%. (From R. K. Shah, D. H. Kim, and D. R. Paul, *Polymer* 48, 1047–1057, 2007. With permission.)

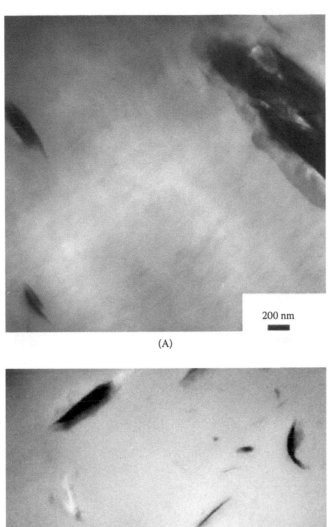

200 nm

(A)

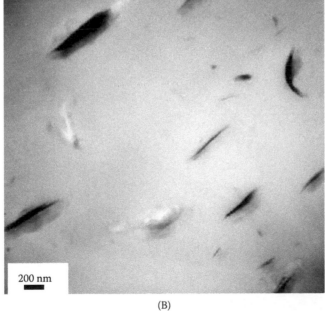

200 nm

(B)

FIGURE 3.50
TEM photomicrographs of nanocomposites prepared from LDPE and the ionomer using a one-tailed organoclay (A, B), M3(HT)1, and a two-tailed organoclay, M2(HT)2 (C, D). The concentration of MMT in all four samples is 2.5 wt%. (From R. K. Shah, D. H. Kim, and D. R. Paul, *Polymer* 48, 1047–1057, 2007. With permission.)

200 nm

(C)

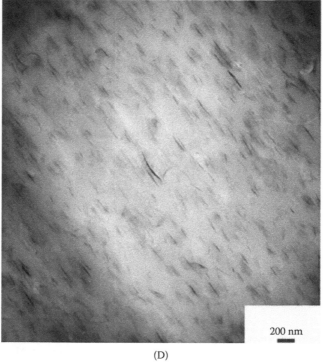

200 nm

(D)

FIGURE 3.50
(*Continued*)

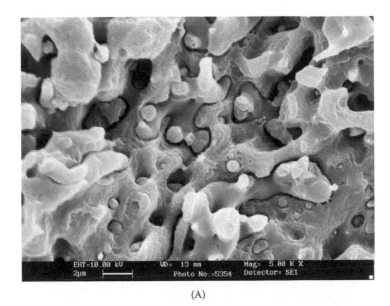

(A)

(B)

FIGURE 3.51

SEM photomicrographs of cryofractured surfaces of melt-mixed blends of polycarbonate with 2 wt% multi-walled carbon nanotubes and polyethylene PC-2NT/PE = 35/65 vol.% illustrating differences in the fineness of cocontinuity when melt-blended at different rotation speeds. (A): screw speed 50 rpm; (B): screw speed 150 rpm. The mixing was carried out using a DACA microcompounder. The cryofracture was performed on extruded strands. (From P. Pötschke, A. R. Bhattacharyya, and A. Janke, *Polymer* 44, 8061–8069, 2003. With permission.)

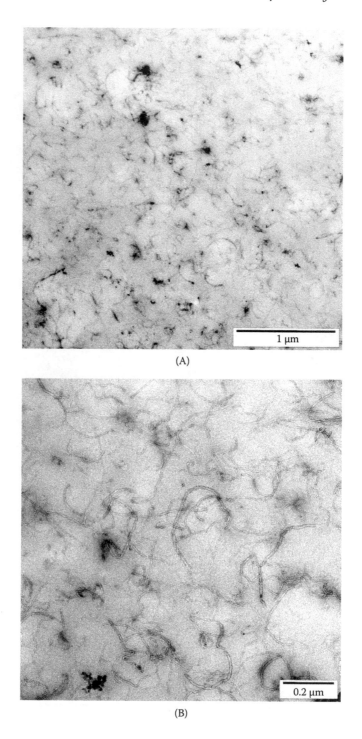

(A)

(B)

FIGURE 3.52

TEM photomicrographs of a composite of polycarbonate with 2 wt% multiwalled carbon nanotubes (MWNT) as obtained by dilution of a master batch with 15 wt% MWNT (delivered by Hyperion Cat. USA) with PC Iupilon E 2000 via melt-mixing using a Haake twin-screw extruder. TEM was made from a granule. (A): low magnification; (B): high magnification. (From P. Pötschke, B. Kretzschmar, and A. Janke, *Compos. Sci. Technol.* 67, 855–860, 2007. With permission.)

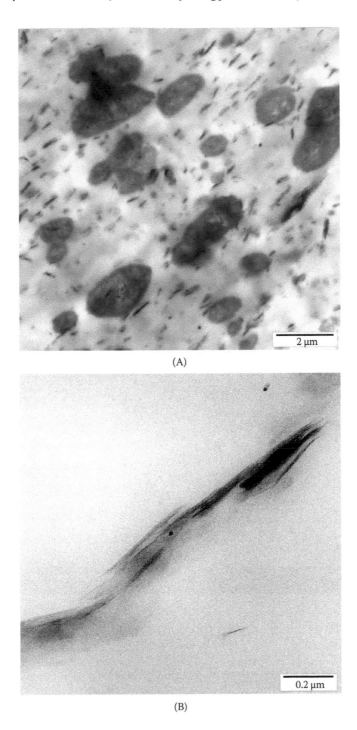

(A)

(B)

FIGURE 3.53

TEM photomicrographs of a composite of polypropylene (two-phase copolymer) with 3 wt% montmorillonite (MMT) Nanofil 15 (DSQ modified from Südchemie, Germany). The material was produced by a two-step procedure using melt-mixing on a ZSK-30. TEM was made from a granule. The image illustrates that MMT is distributed in both phases of the PP and that the clay is not fully exfoliated and some stacks remain in the PP. (A): low magnification; (B): high magnification. (From P. Pötschke, B. Kretzschmar, and A. Janke, *Compos. Sci. Technol.* 67, 855–860, 2007. With permission.)

(A)

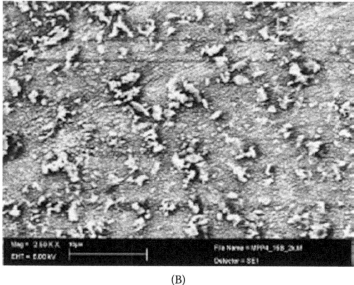

(B)

FIGURE 3.54
SEM of plasma-etched surfaces of PP/MMT/PPy = 95.2/4.8/13.2 (by wt.) composite films (bar size = 10 μm). (A): prepared in water/methanol and processed by direct compression molding; (B): prepared in water/methanol and processed by compression molding of melt-mixed samples; (C): prepared in water/DBSA and processed by direct compression molding; and (D): prepared in water/DBSA and processed by compression molding of melt-mixed samples. (From M. Mravčáková, M. Omastová, P. Pötschke, A. Pozsgay, B. Pukánszky, and J. Pionteck, *Polym. Adv. Technol.* 17, 715–726, 2006. With permission.)

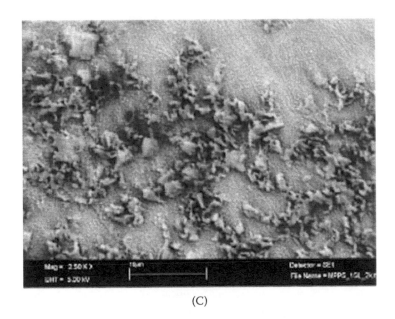

(C)

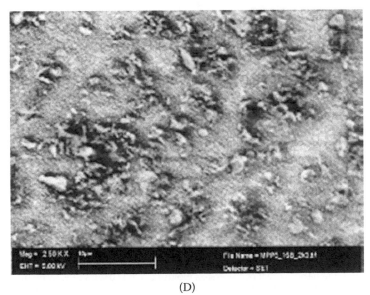

(D)

FIGURE 3.54
(*Continued*)

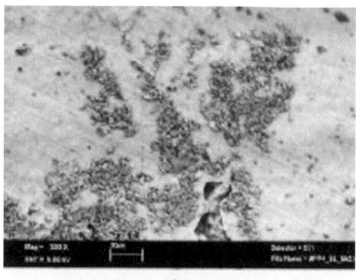

(A)

(B)

FIGURE 3.55
SEM of plasma-etched surfaces of PP/MMT clays modified with polypyrrole PPy (bar size = 20 μm). (A) and (B): 4.8 wt% PPy, water/methanol, direct compression molding; (C): 13.0 wt% PPy, water/methanol, direct compression molding; (D) and (E): 4.8 wt% PPy, water/DBSA, direct compression molding; and (F): 13.0 wt% PPy, water/DBSA, compression molding of melt-mixed sample. Dispersion of montmorillonite clays (MMT) modified with polypyrrole (PPy) in polypropylene (PP) in dependence on processing conditions. (From M. Mravčáková, M. Omastová, P. Pötschke, A. Pozsgay, B. Pukánszky, and J. Pionteck, *Polym. Adv. Technol.* 17, 715–726, 2007. With permission.)

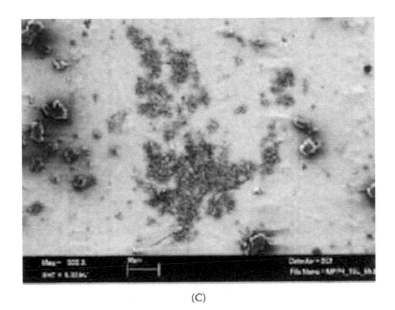

(C)

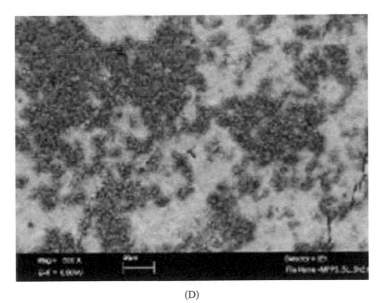

(D)

FIGURE 3.55
(*Continued*)

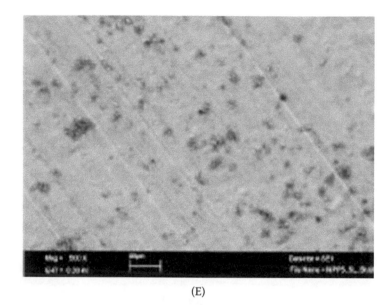

(E)

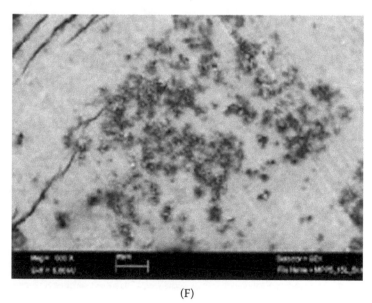

(F)

FIGURE 3.55
(*Continued*)

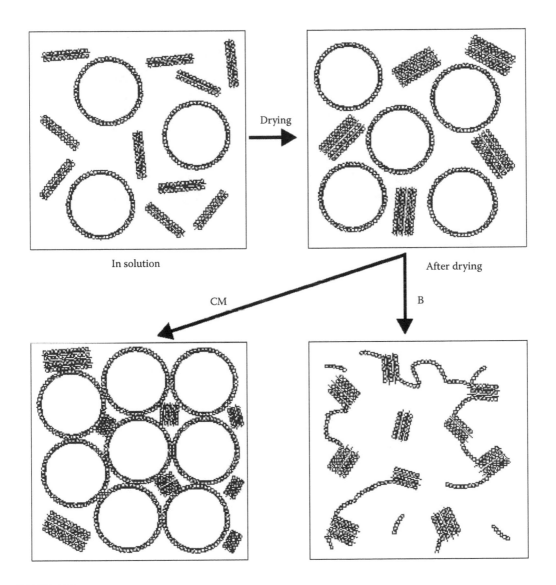

FIGURE 3.56
Scheme of the development of conductive paths in PP/MMT/PPy composites during compression molding (CM) or melt-mixing (B) in connection with Figure 3.54 and Figure 3.55. (From M. Mravčáková, M. Omastová, P. Pötschke, A. Pozsgay, B. Pukánszky, and J. Pionteck, *Polym. Adv. Technol.* 17, 715–726, 2007. With permission.)

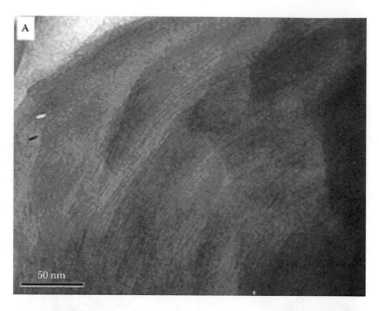

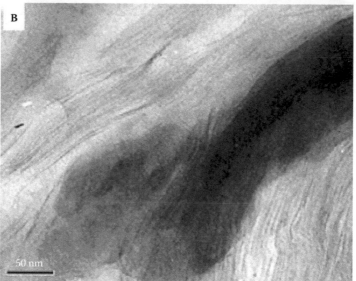

FIGURE 3.57

TEM analysis of sulfonated syndiotactic polystyrene ionomer/organoclay nanocomposites. The structure of the sulfonated syndiotactic polystyrenes (SsPS) ionomer/organoclay nanocomposites with various cations was studied by TEM analysis. The typical TEM photomicrographs of SsPS ionomer/organoclay nanocomposites with an ionomer content of 3.8 mole% with various cations are shown. The parallel lines observed in the photomicrograph of (A) and (B) indicate that intercalated nanocomposites are obtained with H+ and Na+ SsPS/organoclay nanocomposites and the distance between the parallel lines in the photomicrographs are comparable with the d-spacing obtained from WAXD. However, the TEM photomicrographs (C) and (D) of SsPS ionomer/organoclay nanocomposite with K+ and Rb+ cations show that the lines for the clay are disordered, revealing that the clay layers are completely delaminated and exfoliated in the polymer matrix. (From P. Govindaiah, S. R. Mallikarjuna, and C. Ramesh, *Macromolecules* 39, 7199–7203, 2001. With permission.)

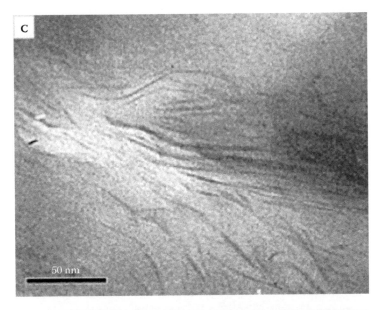

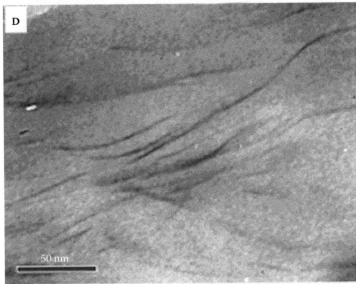

FIGURE 3.57
(*Continued*)

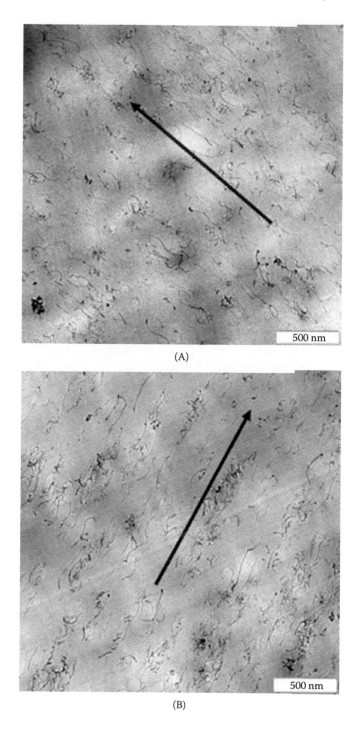

FIGURE 3.58
Orientation of multiwalled carbon nanotubes (MWNT) in melt-spun fibers produced from a melt-compounded polycarbonate composite with 2 wt% MWNT using different take-up velocities: TEM photomicrographs of thin sections of the fibers prepared along the fiber axis (assigned by the arrows). In all images the cut direction is 35° to the fiber axis. Varied is the take-up velocity (drawdown ratio, ddr, in brackets). (A): 50 m/min (ddr = 7.8); (B): 200 m/min (ddr = 31.4); (C): 400 m/min (ddr = 62.8); and (D): 800 m/min (ddr = 126). (From P. Pötschke, H. Brünig, A. Janke, D. Fischer, and D. Jehnichen, *Polymer* 46, 10355–10363, 2005. With permission.)

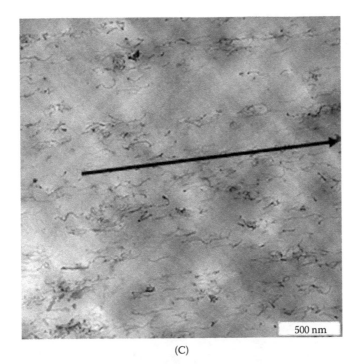

(C)

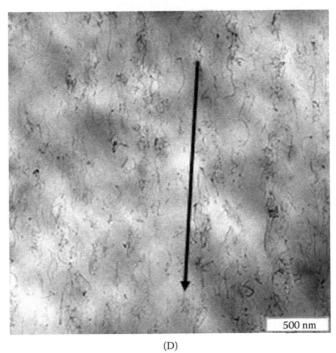

(D)

FIGURE 3.58
(*Continued*)

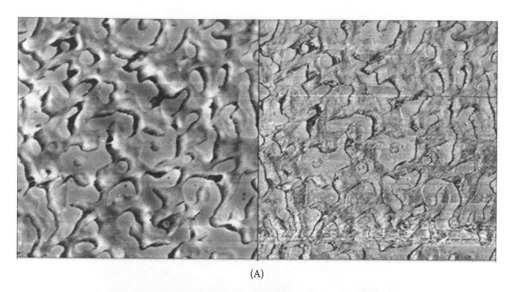

(A)

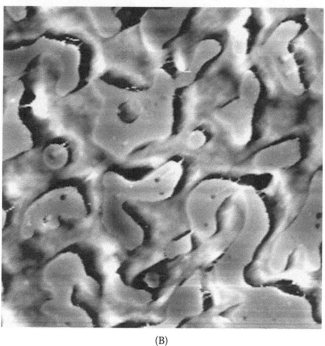

(B)

FIGURE 3.59
Atomic force microscopy (AFM) photomicrographs of a cut surface of a melt-mixed blend of polycarbonate with 2 wt% multiwalled carbon nanotubes (MWNT) and polyethylene PC-2NT/PE = 45/55 vol.% illustrating the cocontinuous structure and the bridging effect of MWNT over the interface due to mismatch between blend phase size and nanotube length. Mixing was done using a DACA microcompounder. Cuts were performed on extruded strands. The frame in (A) image is 12 mm × 12 mm. (From P. Pötschke, A. R. Bhattacharyya, and A. Janke, *Polymer* 44, 8061–8069, 2003. With permission.)

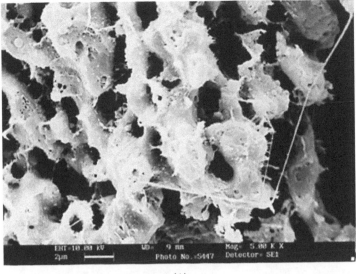

(A)

(B)

FIGURE 3.60

SEM photomicrographs of cryofractured surfaces (extracted in chloroform) of a melt-mixed blend of poly-carbonate with 2 wt% multiwalled carbon nanotubes (MWNT) and polyethylene PC-2NT/PE = 45/55 vol.% (A) and 80/20 vol.% (B) illustrating the bridging effect of MWNT over the interface due to mismatch between blend phase size and nanotube length. Mixing was done using a DACA microcompounder. Cryofractures were performed on extruded strands. (From P. Pötschke, A. R. Bhattacharyya, and A. Janke, *Polymer* 44, 8061–8069, 2003. With permission.)

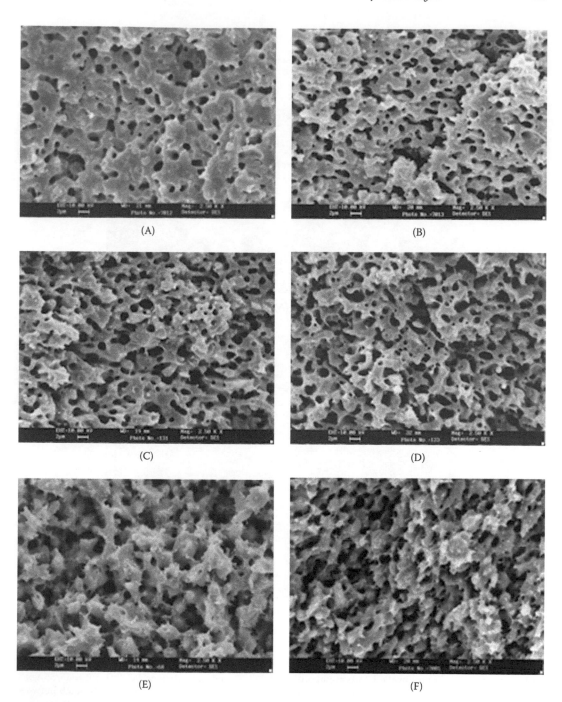

(A)

(B)

(C)

(D)

(E)

(F)

FIGURE 3.61

SEM photomicrographs of cryofractured surfaces (extracted in chloroform) of a melt-mixed blends of poly-carbonate with 2 wt% multiwalled carbon nanotubes (MWNT) and polyethylene illustrating the morphology development over composition. (A): 20PC/80PE; (B): 30/70; (C): 35/65; (D): 40/60; (E): 45/65; (F): 50/50; (G): 60/40; (H): 70/30; and (I): 80/20. Mixing was done using a DACA microcompounder. Cryofractures were performed on extruded strands. (From P. Pötschke, A. R. Bhattacharyya, and A. Janke, *Polymer* 44, 8061–8069, 2003. With permission.)

(G)

(H)

(I)

FIGURE 3.61
(*Continued*)

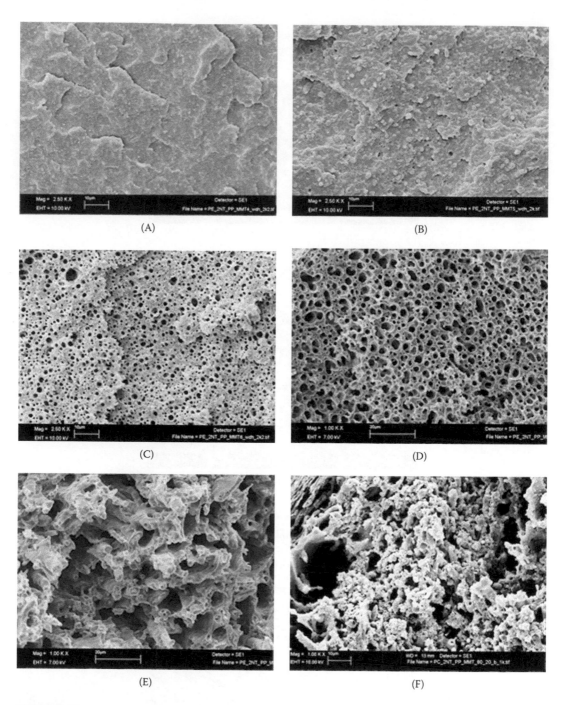

FIGURE 3.62
SEM photomicrographs of cryofractured strands (extracted in chloroform) of different PP-3% MMT/PC-2% carbon nanotube (CNT) (wt%) blends. In 90/10 (A) and 70/30 (B) blends the solvent could not reach the buried dispersed PC-2NT particles. In all other blends—(C): 60/40; (D): 40/60; (E): 30/70; and (F): 20/80—PC forms a continuous structure—and thus is reachable by the solvent—and the remaining PP phase is seen. (From P. Pötschke, B. Kretzschmar, and A. Janke, _Compos. Sci. Technol._ 67, 855–860, 2007. With permission.)

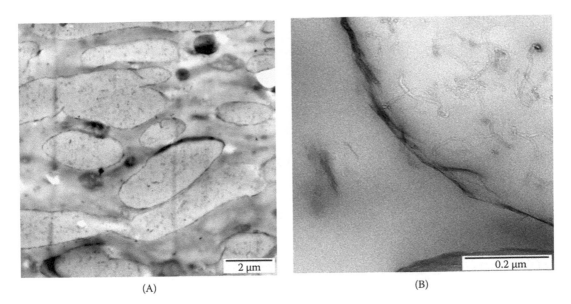

(A) (B)

FIGURE 3.63

TEM photomicrographs of a PP-3% MMT/PC-2% CNT = 60/40 wt% blend. The TEM illustrates that the carbon nanotubes (CNT) are exclusively located and nicely dispersed in the polycarbonate phase (A) and most of the MMT nanoclay is located at the interface between PP and PC in this cocontinuous blend (B). Thus, a nanoclay migration from the prefilled PP phase could be observed during melt-mixing of the blends. (From P. Pötschke, B. Kretzschmar, and A. Janke, *Compos. Sci. Technol. 67*, 855–860, 2007. With permission.)

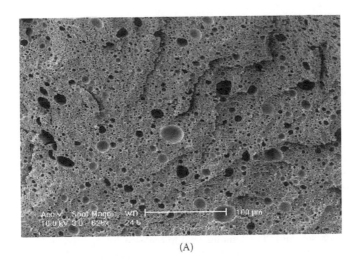

(A)

FIGURE 3.64

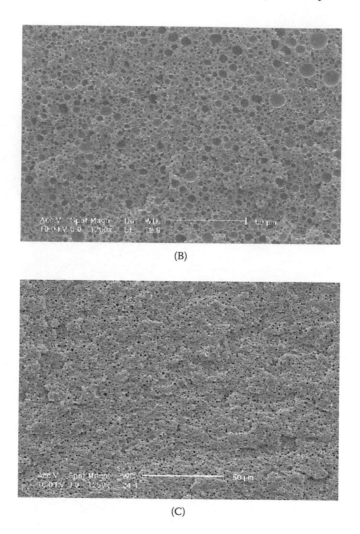

(B)

(C)

FIGURE 3.64 (*Continued*)

SEM photomicrographs of polyphenylene oxide PPO/polyamide 6 PA6 (50/50 w/w) blend with (A): 0% clay; (B): 1% clay; and (C): 2% clay. The organoclay is the synthetic fluorine mica provided by CO-OP chemical. The synthetic mica was selected for this work because of its high aspect ratio. The length of the mica platelets is about 200 to 300 nm and their thickness is about 1 nm. The clay was organically treated with (R-NH3)+Cl⁻ (dipolyoxyethylene alkyl([coco])methylammonium cation) (CEC = 120 mequiv/100 g clay). PA6 and PPO in 50/50 (w/w) proportion were mixed with various amounts of organoclay at 270°C and 100 rpm for 15 min in a twin-screw minimixer (Haake Minilab). Before melt-mixing, all polymers and the clay were dried in a vacuum oven at 80°C for 24 h. PA6 and PPO nanocomposites were also prepared by the same method. The black domains indicate the extracted PPO phase. For the blend without clay, PPO spherical domains are dispersed in the PA6 matrix because of the much higher melt viscosity of PPO than PA6. It is shown from (A) that the PPO domain size distribution is large and the average domain size is about 4.2 μm, which is due to the poor compatibility between PPO and PA6 phases, and PPO domains tend to coalesce during mixing. However, the PPO domain size is decreased to about 3.0 μm upon the addition of only 1% organoclay, and the PPO particles are also uniformly dispersed in the PA6 matrix. The addition of 2% clay causes a further reduction in PPO domain size to about 1.1 μm. (From Y. Li and H. Shimizu, *Polymer* 45, 7381–7388, 2004. With permission.)

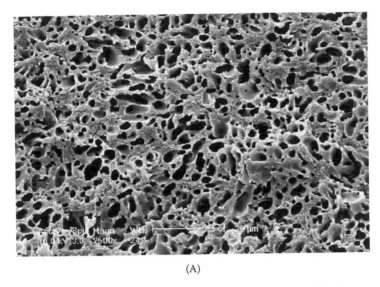

(A)

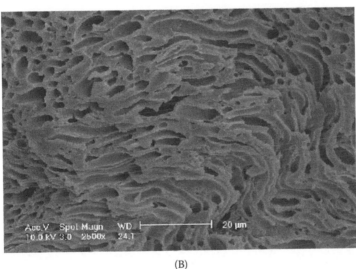

(B)

FIGURE 3.65

SEM photomicrographs of 50 PPO/50 PA6 containing 5 wt% organically modified clay. With further increase of the organoclay content, the blend morphology shows significant change. The SEM images of the fractured surface in directions both perpendicular (A) and parallel (B) to the extrusion direction are displayed. A difference in dispersion type exists between the two directions. An elongated network structure is observed in the direction parallel to extrusion, whereas in the perpendicular direction the black PPO phase appears to be dispersed within the PA6 matrix with many connections between the domains. The morphology is the typical oriented cocontinuous structure, which is elongated along the extrusion direction. This oriented cocontinuous structure can transform into an isotropic structure when annealed at 270°C for 15 min, as shown in (C). Both PPO and PA6 phases have isotropic three-dimensional spatial continuity in the nanocomposite. (From Y. Li and H. Shimizu, *Polymer* 45, 7381–7388, 2004. With permission.)

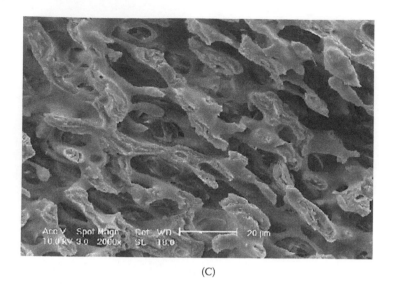

(C)

FIGURE 3.65
(*Continued*)

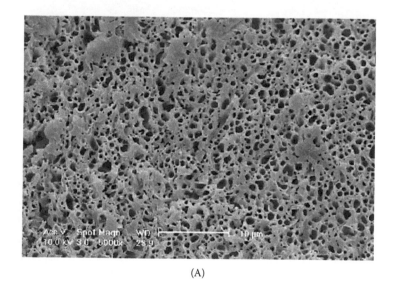

(A)

FIGURE 3.66
SEM micrographs of a PPO/PA6 blend nanocomposite with (A) 10% clay and (B) 20% clay. The cocontinuous morphology becomes finer with higher organoclay content, as clearly shown in this figure in contrast to Figure 3.64 and Figure 3.65. The figure also shows the oriented cocontinuous structure for the nanocomposites with 10 and 20% clay. When the clay content is increased, a significant decrease in phase sizes is observed. The phase size is about 1.2 μm for the nanocomposite with the addition of 10% clay, but for the nanocomposite with 20% clay, the phase size is significantly decreased to 0.5 μm. (From Y. Li and H. Shimizu, *Polymer* 45, 7381–7388, 2004. With permission.)

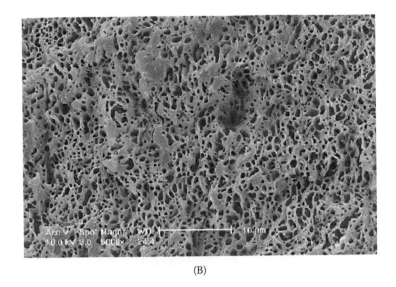

(B)

FIGURE 3.66
(*Continued*)

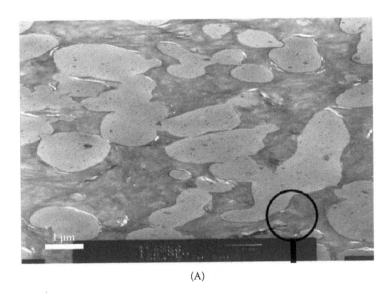

(A)

FIGURE 3.67
TEM micrographs of a PPO/PA6 nanocomposite with 5% clay (A): low magnification; (B): high magnification of the circled area. The white part with small domains is the PPO phase, and the other part is the PA6 phase. The dark line in the PA phase is the organoclay. These images show that the clay locates only in the PA6 phase. This indicates that the exfoliated clay platelets will only change the rheological properties of PA6 but will not affect those of PPO. (From Y. Li and H. Shimizu, *Polymer* 45, 7381–7388, 2004. With permission.)

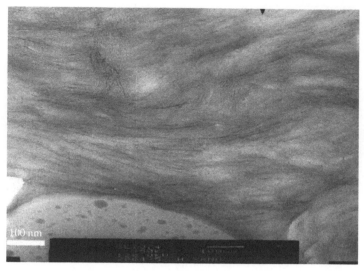

(B)

FIGURE 3.67
(*Continued*)

References

1. Kojima, Y., A. Usuki, M. Kawasumi, A. Okada, Y. Fukushima, T. Kurauchi, and O. Kamigaito, *J. Mater. Res.*, 8, 1185 (1993).
2. Kojima, Y., A. Usuki, M. Kawasumi, A. Okada, T. Kurauchi, and O. Kamigaito, *J. Appl. Polym. Sci.*, 49, 1259 (1993).
3. Kojima, Y., K. Fukumori, A. Usuki, A. Okada, and T. Kurauchi, *J. Mater. Sci. Lett.*, 12, 889 (1993).
4. Usuki, A., M. Kawasumi, Y. Kojima, and A. Okada, *J. Mater. Res.*, 8, 1174 (1993).
5. Usuki, A., Y. Kojima, M. Kawasumi, A. Okada, Y. Fukushima, and T. Kurauchi, *J. Mater. Res.*, 8, 1179 (1993).
6. Giannelis, E. P., *Adv. Mater.*, 8, 29 (1996).
7. Giannelis, E. P., R. Krishnamoorti, and E. Manias, *Adv. Polym. Sci.*, 138, 107 (1999).
8. Vaia, R. A., G. Price, P. N. Ruth, H. T. Nguyen, and J. Lichtenhan, *Appl. Clay Sci.*, 15, 67 (1999).
9. Giannelis, E. P., *Appl. Organomet. Chem.*, 12, 675 (1998).
10. Bharadwaj, R. K., *Macromolecules*, 34, 1989 (2001).
11. Messersmith, P. B. and E. P. Giannelis, *J. Polym. Sci., Part A: Polym. Chem.*, 33, 1047 (1995).
12. Powell, C. E. and G. W. Beall, *Curr. Op. Solid. St. Mater. Sci.*, 10, 73 (2006).
13. Gilman, J. W., T. Kashiwagi, and J. D. Lichtenhan, *SAMPE J.*, 33, 40 (1997).
14. Gilman, J. W., *Appl. Clay Sci.*, 15, 49 (1999).
15. Vaia, R. A., H. Ishii, and E. P. Giannelis, *Chem. Mater.*, 5, 1694 (1993).
16. Suprakas, S. R. and M. Okamoto, *Prog. Polym. Sci.*, 28, 1539 (2003).
17. Liu, L., Z. Qi, and X. Zhu, *J. Appl. Polym. Sci.*, 71, 1133 (1999).
18. Usuki, A., M. Kato, A. Okata, and T. Kurauchi, *J. Appl. Polym. Sci.*, 63, 137 (1997).
19. Nam, P. H., P. Maiti, M. Okamoto, T. Kotaba, N. Hasegawa, and A. Usuki, *Polymer*, 42, 9633 (2001).
20. Reichert, P., B. Hoffmann, T. Bock, R. Thomann, R. Mülhaupt, and C. Friedrich, *Macromol. Rapid. Comm.*, 22, 519 (2001).

21. Hambir, S., N. Bulakh, P. Kodgire, R. Kalgaonkar, and J. P. Jog, *J. Polym. Sci. Part B: Polym. Phys.*, 39, 446 (2001).
22. Liu, X. and Q. Wu, *Polymer*, 42, 10013 (2001).
23. Tjong, S. C., Y. Z. Meng, and A. S. Hay, *Chem. Mater.*, 14, 44 (2002).
24. Sun, T. and J. M. Garces, *Adv. Mater.*, 14, 128 (2002).
25. Okada, A., M. Kawasumi, T. Kurauchi, and O. Kamigaito, *ACS Polym. Prep.*, 28, 447 (1987).
26. Kojima, Y., *J. Polym. Sci. Part A: Polym. Chem.*, 31, 1755 (1993).
27. Messersmith, P. B. and E. P. Giannelis, *Chem. Mater.*, 6, 468 (1994).
28. Wang, M. S. and T. J. Pinnavaia, *Chem. Mater.*, 5, 1064 (1994).
29. Dong Choo Lee and Lee Wook Jang, *J. Appl. Polym. Sci.*, 61, 1117 (1996).
30. Biasci, L. et al., *Polymer*, 35, 3296 (1994).
31. Messersmith, P. B. and E. P. Giannelis, *Chem. Mater.*, 5, 1064 (1993).
32. Zilg, C., R. Thomas, R. Mülhaupt, and J. Finter, *Adv. Mater.*, 11, 49 (1999).
33. Yano, K., A. Usuki, and A. Okada, *J. Polym. Sci. Part A: Polym. Chem.*, 35, 2289 (1997).
34. Lan, T., P. D. Kaviratna, and T. J. Pinnavaia, *Chem. Mater.*, 6, 573 (1994).
35. Reichert, P., J. Kressler, R. Thomann, R. Mülhaupt, and G. Stöpperlmann, *Acta Polym.*, 49, 116 (1998).
36. Kornmann, X., L. A. Berglund, and J. Sterte, *Polym. Eng. Sci.*, 38, 1351 (1998).
37. Matayabas, J. C. and S. R. Turner, Patent WO 98/29499 Eastman Chemicals.
38. Korbee, A., Patent Appl. WO 99/29767 DSM.
39. Kato, M., A. Usuki, and A. Okada, *J. Appl. Polym. Sci.*, 66, 1781 (1997).
40. Usuki, A., M. Kato, and A. Okada, *J. Appl. Polym. Sci.*, 67, 137 (1997).
41. Kushta, F. D., P. J. Lemstra, A. Keller, L. F. Batenburg, and L. F. Fischer, *MRS Proceedings*, San Francisco, CA (April 1999).
42. Goettler, L. A. and D. W. Recktenwald, *Proccedings of Additives 98*, Orlando, FL (1998).
43. Kurukawa, Y., H. Yasuda, and A. Oya, *J. Mater. Sci. Letters*, 15, 1481 (1996).
44. Jeon, H. G., H. T. Jung, S. W. Lee, and S. D. Hudson, *Polym. Bull.*, 41, 17 (1998).
45. Beall, G. W., A. Sorokin, S. Tsipursky, and A. Goldman, Patent EP0747451A2.
46. Ogata, N., S. Kawakage, and T. Ogihara, *J. Appl. Polym. Sci.*, 66, 573 (1997).
47. Hui, C. Y. and D. Shia, *Polym. Eng. Sci.*, 38, 774 (1998).
48. Shelly, J. S., P. T. Mather, and K. L. DeVries, *Polymer*, 42, 5849 (2001).
49. Shen, Z., G. P. Simon, and Y.-B. Cheng, *Polymer*, 43, 4251 (2002).
50. Li, Y. and H. Ishida, *ACS Polym. Mater. Sci. Eng.*, Fall Meeting, Vol. 85 (2001).
51. Sukpirom, N. and M. M. Lerner, *Chem. Mater.*, 13, 2179 (2001).
52. Aranda, P. and E. Ruiz-Hitzky, *Chem. Mater.*, 4, 1395 (1992).
53. Gendoya, I., D. Lopez, A. Alegria, and C. Mijangos, *J. Polym. Sci.: Polym. Phys. Ed.*, 39, 1968 (2001).
54. Greenland, D. J., *J. Colloid Sci.*, 18, 647 (1963).
55. Krishnamoorti, R., R. A. Vaia, and E. P. Giannelis, *Chem. Mater.*, 8, 1728 (1996).
56. Blumstein, A., *J. Polym. Sci. Part A: Polym. Chem.*, 3, 2653 (1965).
57. Manias, E., H. Chen, R. Krishnamoorti, K. Genzer, and E. P. Giannelis, *Macromolecules*, 33, 7955 (2000).
58. Maine, F. W. and P. D. Shepherd, *Composites*, 193–200 (1974).

4

Illustrations of Phase Morphology Other Than Cocontinuous in Polymer Blends Including Thermoplastics/Thermoplastics, Thermosets/Thermoplastics

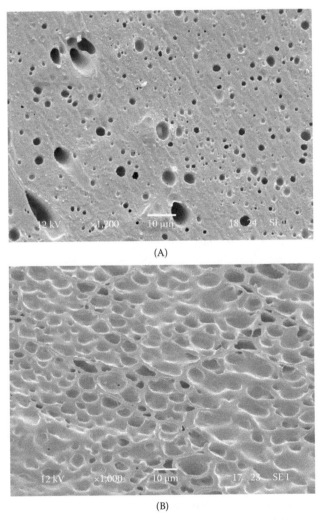

(A)

(B)

FIGURE 4.1
Scanning electron microscopy (SEM) photomicrographs of very-low-density polyethylene (VLDPE)/styrene-ethylene butylene-graft-diethyl maleate (SEBS-g-DEM)/polyethylene terephthalate (PET) 40/40/20 blend cryogenic fracture. (A): after extraction with 1,1,1, 3,3,3 hexafluoro 2 propanol (HFIP); (B): after the successive extraction with chloroform. (From M. B. Coltelli, I. Della Maggiore, S. Savi, M. Aglietto, and F. Ciardelli, *Polym. Degrad. Stab.* 90(2), 211–223, 2005; erratum 91, 987, 2006. With permission.)

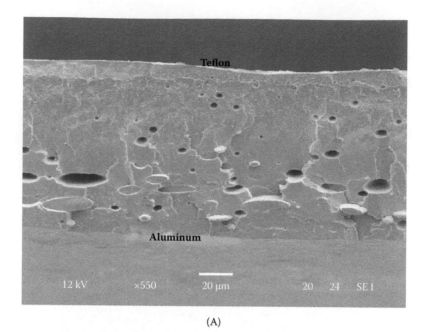

(A)

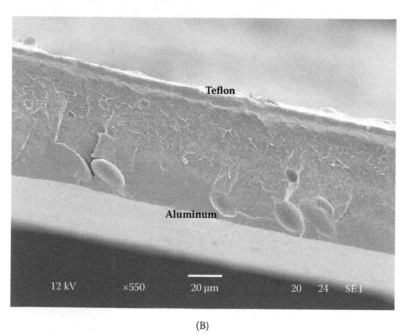

(B)

FIGURE 4.2
SEM photomicrographs of the cryogenic fracture surface of a film made of 90/10 polyethylene/copolyamide (PE/CPA) blend prepared via compression molding at 180°C and 125 bar for 10 min (A), 30 min (B), and 60 min (C). (From M. Bertoldo, M. B. Coltelli, L. Miraglia, P. Narducci, and S. Bronco, *Polymer* 46, 11311–11321, 2005. With permission.)

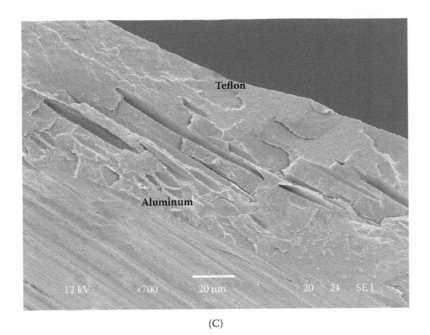

(C)

FIGURE 4.2
(*Continued*)

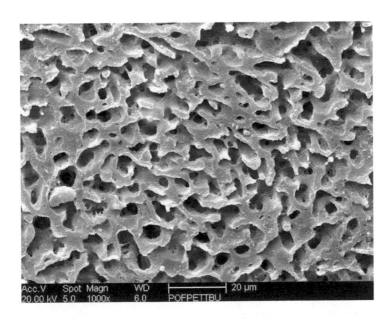

FIGURE 4.3
SEM photomicrograph of the smoothed and toluene-etched surface of a PET ultra-low-density polyethylene (ULDPE)-g-DEM blend obtained in a discontinuous mixer by adding Ti(OBu)4 as a transesterification catalyst. (From M. B. Coltelli, *Catalysed Reactive Compatibilization of Polyolefin and Poly(ethylene terephthalate) Blends: Reactions Mechanisms and Phase Morphology Development*, Ph.D. thesis, University of Pisa, Italy, 2005.)

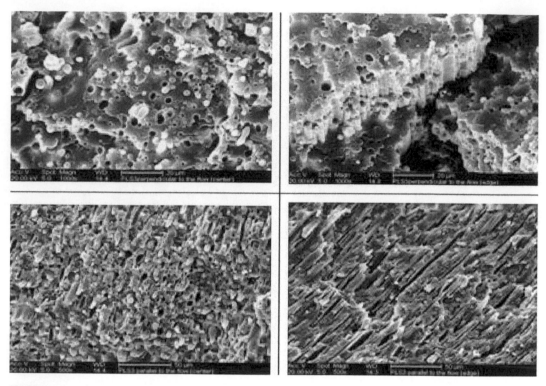

FIGURE 4.4

SEM photomicrographs of PET/LDPE/SEBS-g-DEM 80/13.3/6.6 blends. Top images: before etching (left) and after etching (right). Bottom images: cryogenically broken surface in the center and near to the edge of the extruded strand. (From M. B. Coltelli, C. Harrats, M. Aglietto, and G. Groeninckx, EuroFillers and Polymer Blends Joint Meeting, Proceedings, p. 19, Brugge, Belgium, 2005.)

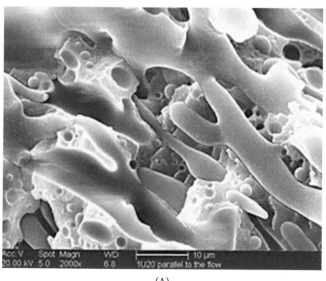

(A)

FIGURE 4.5

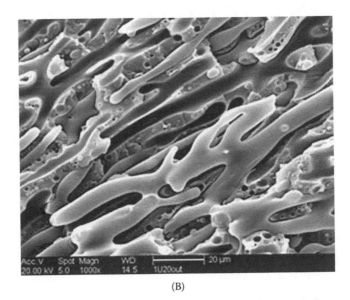

(B)

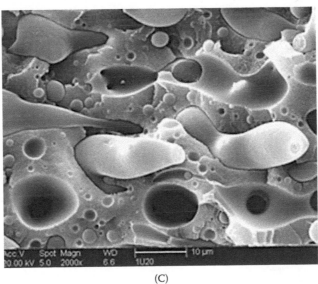

(C)

FIGURE 4.5 *(Continued)*
SEM photomicrographs of the 70/30 PET/LDPE+diethyl maleate grafted ultra-low-density polyethylene (ULDPE-g-DEM) blends obtained with a mixing time of 20 min. (A) and (B): no copolymer, sample cut parallel to flow direction; (C): no copolymer, sample cut perpendicular to flow direction. (From M. B. Coltelli, *Catalysed Reactive Compatibilization of Polyolefin and Poly(ethylene terephthalate) Blends: Reactions Mechanisms and Phase Morphology Development,* Ph.D. thesis, University of Pisa, Italy, 2005.)

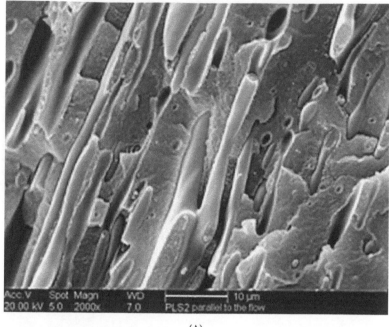

(A)

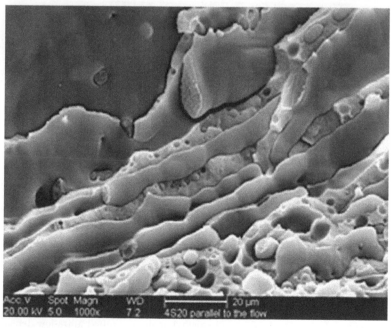

(B)

FIGURE 4.6

SEM photomicrographs of PET/LDPE+SEBS-g-DEM blends obtained using a mixing time of 20 min. (A): 80/16.7 PET/LDPE containing 3.3 wt% of SEBS-g-DEM sample cut parallel to flow direction; (B): 70/20 PET/LDPE blend containing 10 wt% of SEBS-g-DEM, sample cut parallel to flow direction. (From M. B. Coltelli, *Catalysed Reactive Compatibilization of Polyolefin and Poly(ethylene terephthalate) Blends: Reactions Mechanisms and Phase Morphology Development,* Ph.D. thesis, University of Pisa, Italy, 2005.)

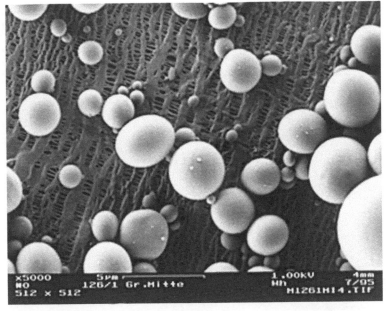

(A)

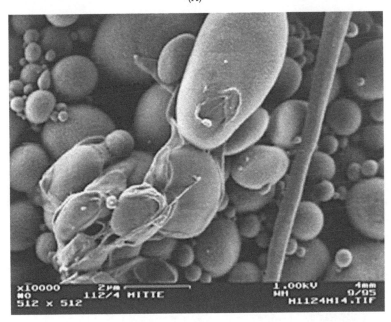

(B)

FIGURE 4.7

Visualization of compatibilizer location in immiscible blends: SEM photomicrographs of a blend of thermoplastic polyurethane (TPU) with 20 wt% polypropylene compatibilized with different ethylenic co- and terpolymers (blend melt-mixed using ZSK-30 extruder). (A): TPU/propylene (PP) = 80/20 wt% blend; (B): added with 5 wt% Lucalen having 4% acrylic acid; (C): added with 5 wt% Lucalen without acrylic acid; (D): added with 5 wt% Luwax having 20% acrylic acid. The TPU matrix was selectively dissolved in dimethylformamide and the remaining PP particles without or with compatibilizer were separated on a membrane. (From K. Wallheinke, W. Heckmann, P. Pötschke, and H. Stutz, *Polym. Test.* 17(4), 247–255, 1998. With permission.)

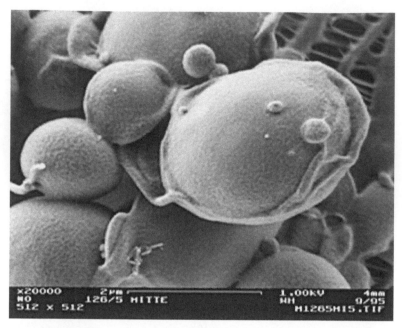

(C)

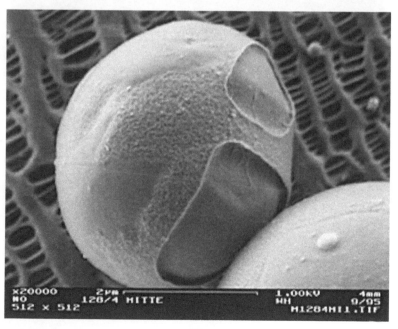

(D)

FIGURE 4.7
(*Continued*)

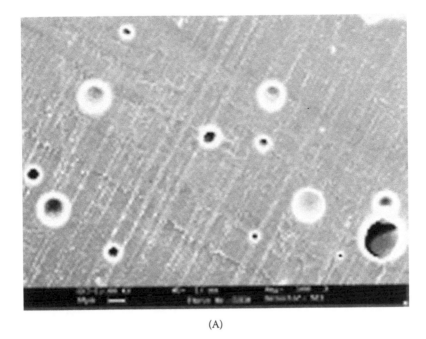

(A)

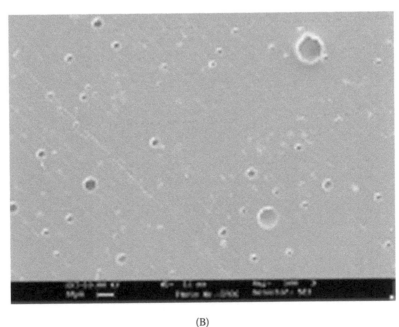

(B)

FIGURE 4.8
SEM photomicrographs of blends based on hyperbranched polymers with a functionality of ca. 100 COOH-groups per molecule as reactive compatibilizers in blends based on oxazoline-terminated PP (PP-Ox) and oxazoline-terminated PS (PS-Ox). Distribution of 5 wt% hyperbranched polymer (HBP) in PP, PP-Ox, PS, and PS-Ox. (A): PP/HBP; (C): PS/HPB nonreactive; (B): PP-Ox/HPB; and (D): PS-Ox/HPB reactive, SEM on cryocuts, HBP etched with THF, frame size 240 × 180 μm. (From J. Pionteck, P. Pötschke, N. Proske, H. Zhao, H. Malz, D. Beyerlein, U. Schulze, and B. Voit, *Macromol. Symp.* 198, 209–220, 2003. With permission.)

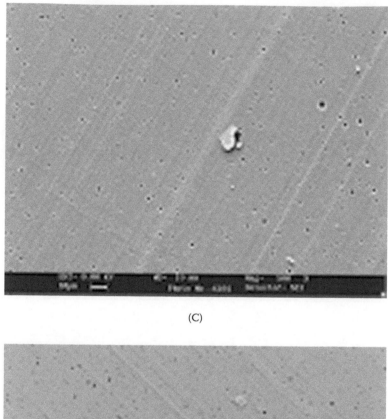

(C)

(D)

FIGURE 4.8
(*Continued*)

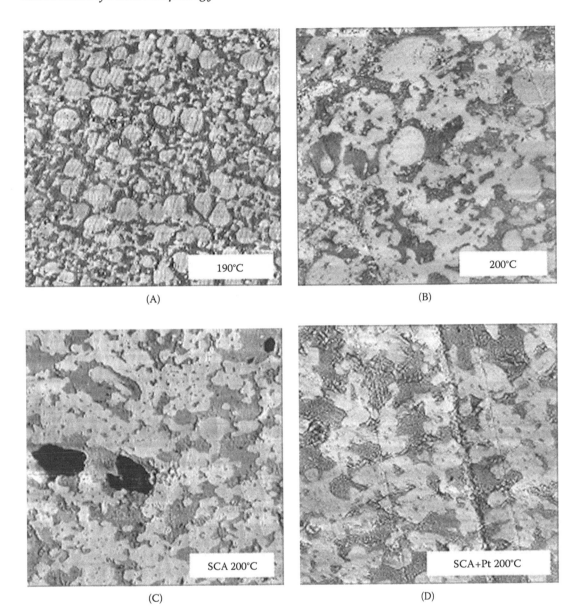

FIGURE 4.9
Atomic force microscopy (AFM) images in phase-contrast mode of PS-COOH/polymethyl methacrylate (PMMA)-NH2 = 40/60 (vol.%) prepared at 190°C (A); (B) like (A), prepared at 200°C; (C) like (B), with SCA; (D) like (B), with SCA+Pt (frame size 10 × 10 μm). Influence of silane-containing coupling agent (SCA) and processing conditions on morphology of PS-COOH/PMMA-NH2 = 40/60 (vol.%) blends. (From J. Pionteck, V. B. Sadhu, L. Jakisch, P. Pötschke, L. Häußler, and A. Janke, *Polymer* 46, 6563–6574, 2005. With permission.)

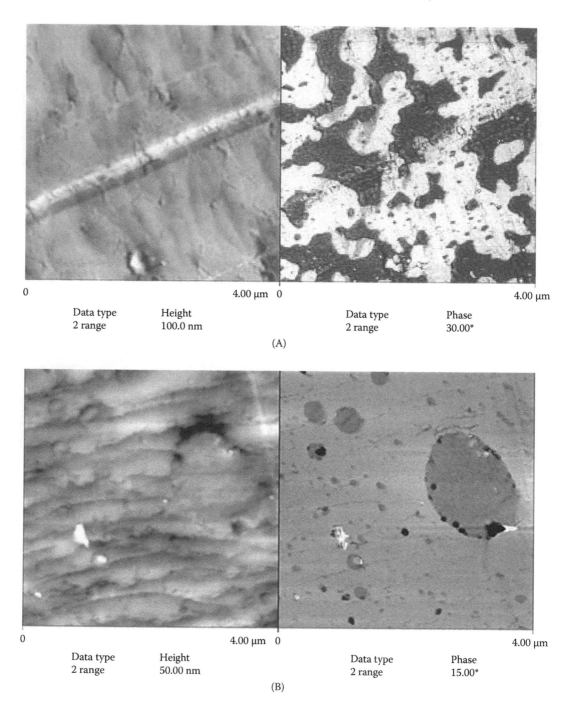

0   4.00 μm 0   4.00 μm

Data type	Height		Data type	Phase
2 range	100.0 nm		2 range	30.00*

(A)

0   4.00 μm 0   4.00 μm

Data type	Height		Data type	Phase
2 range	50.00 nm		2 range	15.00*

(B)

FIGURE 4.10
AFM analysis of (A) PS-COOH/PMMA-NH2/SCA/Pt, 40/60/3, melt-mixed at 200°C for 30 min; (B) PS-COOH/PMMA-NH2/SCA, 40/60/3 melt-mixed at 200°C for 10 min and annealed at 200°C for 30 min, without shear (left images: height profile; right images: stiffness contrast; frame size 4 × 4 μm). Influence of silane-containing coupling agent (SCA) and processing conditions on morphology of PS-COOH/PMMA-NH2 = 40/60 (vol.%) blends. (From J. Pionteck, V. B. Sadhu, L. Jakisch, P. Pötschke, L. Häußler, and A. Janke, *Polymer* 46(17), 6563–6574, 2005. With permission.)

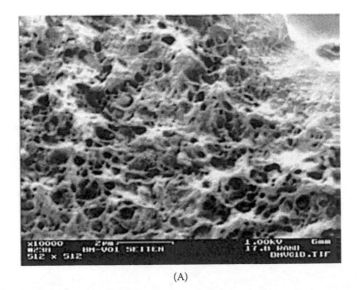

(A)

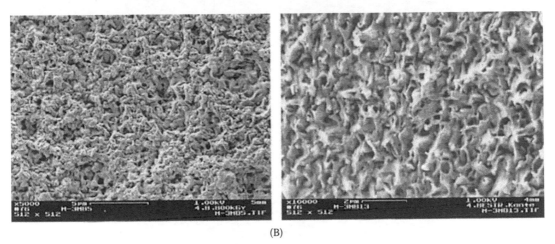

(B)

FIGURE 4.11

Porous structures are formed from Interpenetrating network (IPN) due to extraction in hot xylene after irradiation with an electron beam (doses in kGy), which destroys the polymethacrylate phase, cross-links the PE phase, and causes partial grafting between phases. The SEM images show cryofractures of different treated samples. MMA: methyl methacrylate; EMA: ethyl methacrylate; BMA: n-butyl methacrylate: DMA: n-dodecyl methacrylate; BDDM: 1,4-butanediol dimethacrylate; peroxide: 2,5-dimethyl-2,5-di(tert-butylperoxy)hexane (Trigonox-101). (A): Porous structure of the cross section of the PE/BMA-co-MMA IPN (polytetrafluoroethylene [PTFE] reactor, BMA/MMA = 50:50, BDDM = 0.1 mol%, peroxide = 1 wt%, 800 kGy). A dense surface structure is visible in the right upper corner. Porous surface (A) and bulk morphology (B) of the PE/DMA-co-EMA IPN (PTFE reactor, no BDDM, peroxide = 3 wt%, 800 kGy). (B): porous (left) surface and (right) bulk morphology of the PE/DMA-co-EMA IPN (PTFE reactor, no BDDM, peroxide = 3 wt%, 800 kGy). (C): porous surface structure of the PE/BMA-co-MMA IPN (Alu reactor, BMA/MMA = 50:50, BDDM = 1 mol%, peroxide = 1 wt%, 800 kGy). (D): porous (left) surface and (right) bulk morphology of the PE/BMA-co-MMA IPN (Alu reactor, BMA/MMA = 60:40, BDDM = 1 mol%, peroxide = 1 wt%, 400 kGy). (E): porous surface structure of the PE/DMA-co-EMA IPN (Alu reactor, BDDM = 1 mol%, peroxide = 1 wt%, 800 kGy). (From J. Pionteck , J. Hu, and U. Schulze, *J. Appl. Polym. Sci.* 89, 1976–1982, 2003. With permission.)

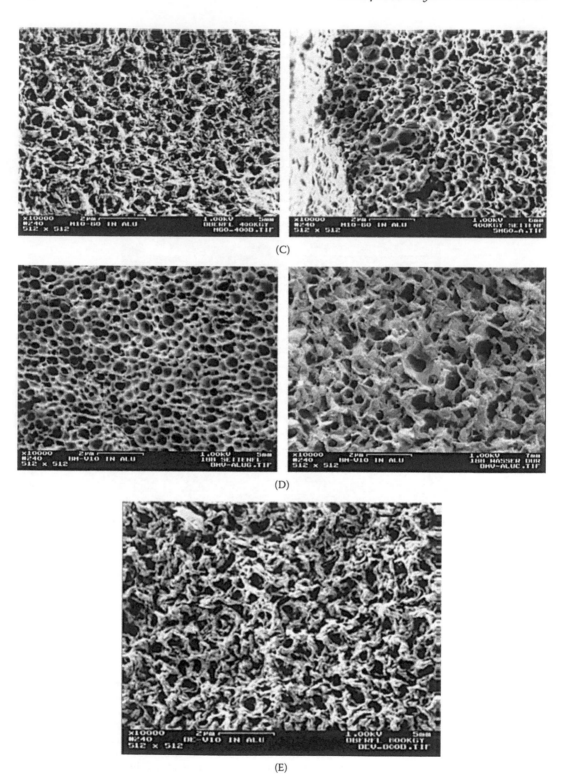

FIGURE 4.11
(*Continued*)

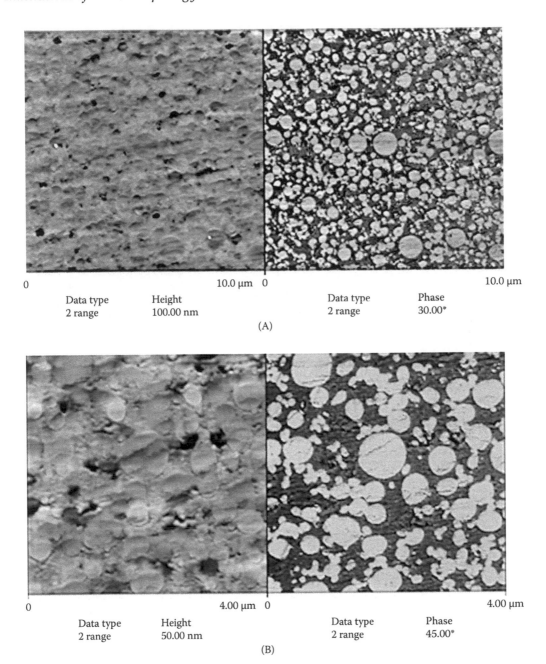

FIGURE 4.12
AFM images of blends of PS-COOH/PMMA-NH2 (PS/PM) compatibilized with an SCA. (A) and (B): PS/PM = 50/50 without compatibilizer. (C) and (D): PS/PM/SCA (50/50/3). (E) and (F): PS/PM/SCA (50/50/3) + catalyst. (G) and (H): PS/PM (40/60) without compatibilizer. (I) and (J): PS/PM (40/60) without compatibilizer after annealing 30 min at 200°C. (K) and (L): PS/PM/SCA (40/60/3). (M) and (N): PS/PM/SCA (40/60/3) after 30 min of annealing at 200°C. (O) and (P): PS/PM/SCA (40/60/3) + catalyst. (Q) and (R): PS/PM/SCA (40/60/3) + catalyst after annealing 30 min at 200°C. NanoScope IV-Dimension 3100 (Veeco) on smooth cut surfaces, tapping mode, topography (left), and phase images (right) detected simultaneously; some images are published in cited paper, composition in volume. For details concerning structure and chemistry, see the reference. (From J. Pionteck, V. B. Sadhu, L. Jakisch, P. Pötschke, L. Häußler, and A. Janke, *Polymer* 46, 6563–6574, 2005. With permission.)

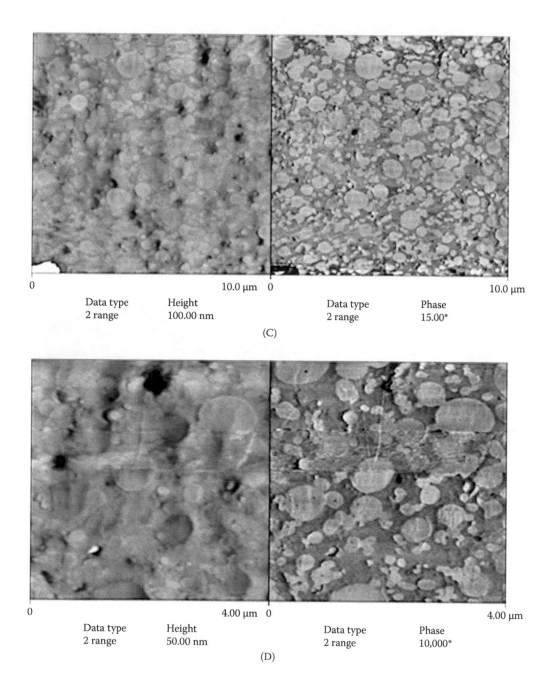

FIGURE 4.12
(*Continued*)

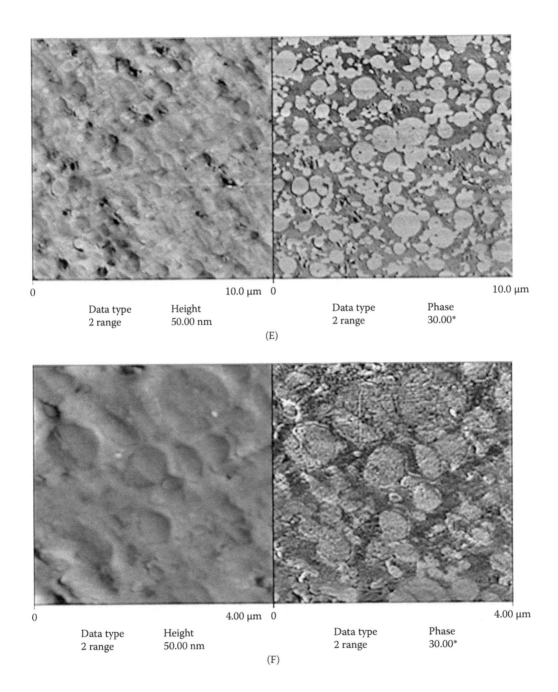

0   10.0 μm   0   10.0 μm

| Data type | Height | | Data type | Phase |
| 2 range | 50.00 nm | | 2 range | 30.00* |

(E)

0   4.00 μm   0   4.00 μm

| Data type | Height | | Data type | Phase |
| 2 range | 50.00 nm | | 2 range | 30.00* |

(F)

FIGURE 4.12
(*Continued*)

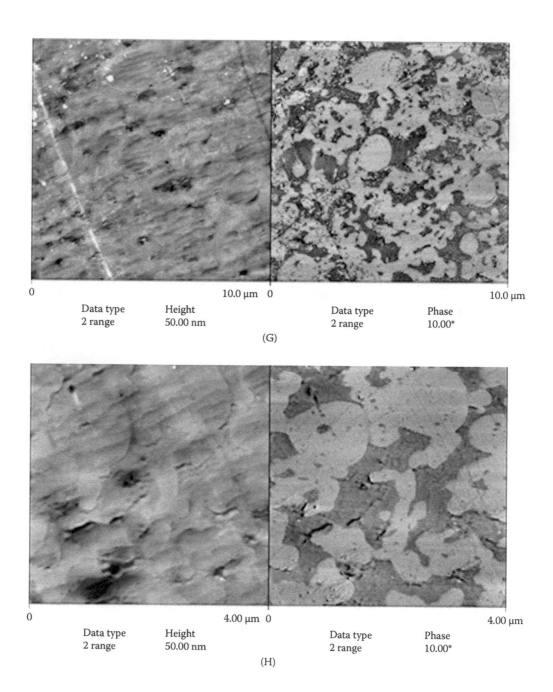

FIGURE 4.12
(*Continued*)

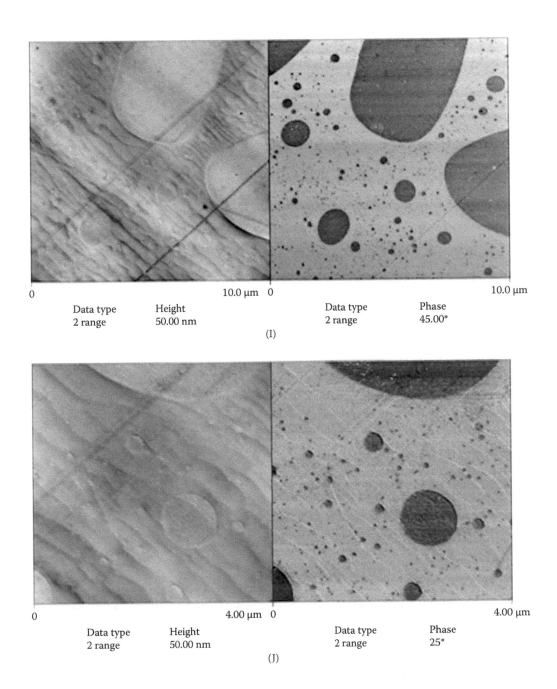

0   10.0 μm 0   10.0 μm

Data type   Height   Data type   Phase
2 range   50.00 nm   2 range   45.00*

(I)

0   4.00 μm 0   4.00 μm

Data type   Height   Data type   Phase
2 range   50.00 nm   2 range   25*

(J)

FIGURE 4.12
(*Continued*)

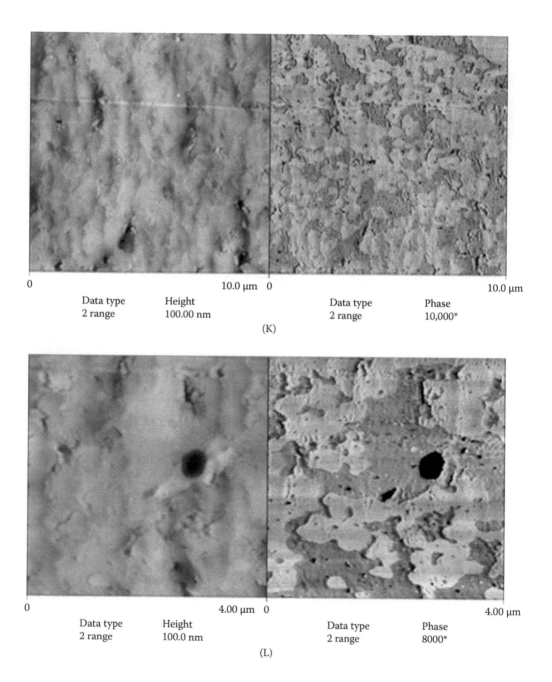

FIGURE 4.12
(*Continued*)

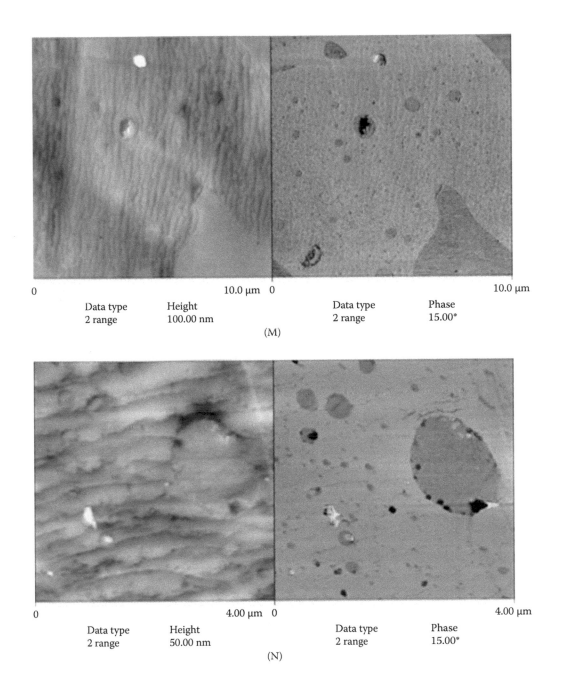

FIGURE 4.12
(*Continued*)

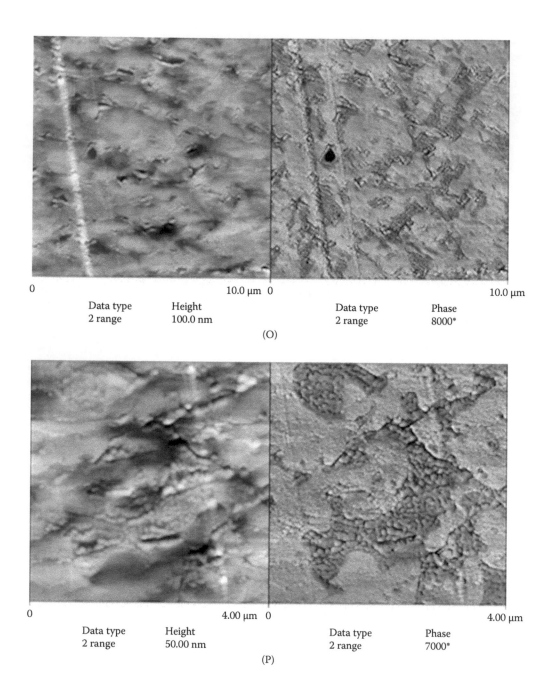

0	10.0 μm	0	10.0 μm
Data type	Height	Data type	Phase
2 range	100.0 nm	2 range	8000°

(O)

0	4.00 μm	0	4.00 μm
Data type	Height	Data type	Phase
2 range	50.00 nm	2 range	7000°

(P)

FIGURE 4.12
(*Continued*)

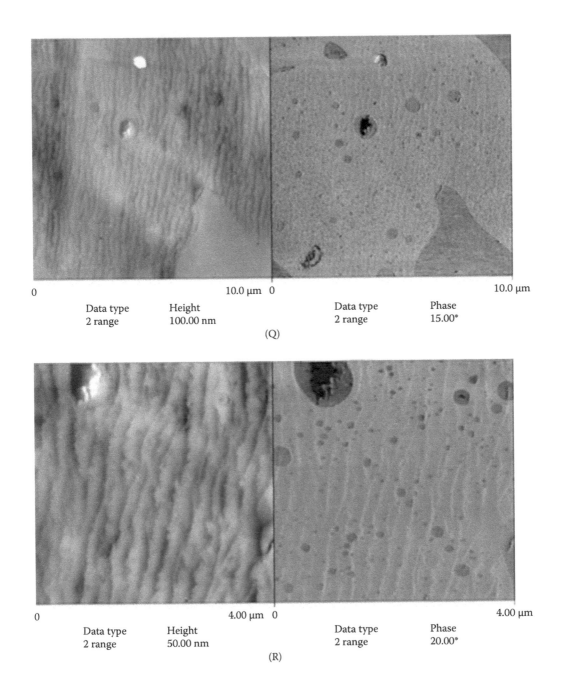

0   10.0 µm 0   10.0 µm

Data type   Height   Data type   Phase
2 range   100.00 nm   2 range   15.00*

(Q)

0   4.00 µm 0   4.00 µm

Data type   Height   Data type   Phase
2 range   50.00 nm   2 range   20.00*

(R)

FIGURE 4.12
(*Continued*)

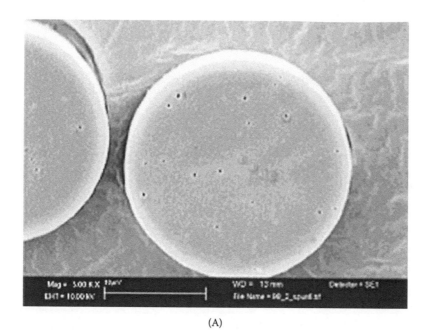

(A)

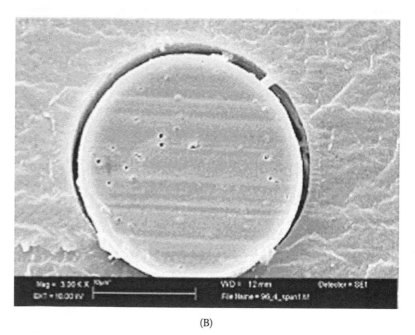

(B)

FIGURE 4.13

Dispersion of PS-fibrils in spun PP/PS composite fibers. SEM photomicrographs of cross sections. (A): PP/PS 98/2; (B): PP/PS 96/4; (C): PP/PS 94/6; (D): PP/PS 92/8. (PS phase is etched out with xylene at room temperature.) (From Q. Xing, M. Zhu, Y. Wang, Y. Chen, Y. Zhang, J. Pionteck, and H. J. Adler, *Polymer* 46(14), 5406–5416, 2005. With permission.)

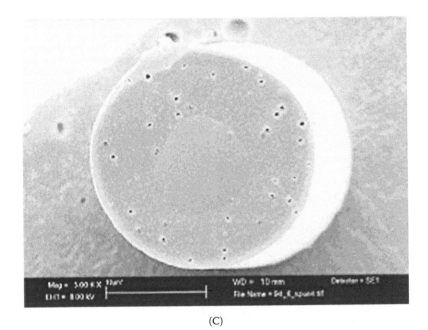

(C)

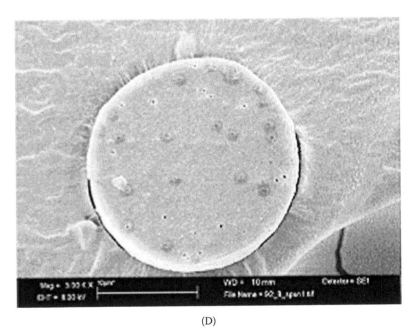

(D)

FIGURE 4.13
(*Continued*)

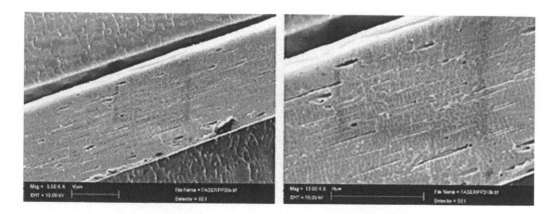

FIGURE 4.14

Dispersion of PS-fibrils in as-spun PP/PS 92/8 composite fibers, SEM photomicrographs in longitudinal section. PS phase is etched out with xylene at room temperature. (From Q. Xing, M. Zhu, Y. Wang, Y. Chen, Y. Zhang, J. Pionteck, and H. J. Adler, *Polymer* 46(14), 5406–5416, 2005. With permission.)

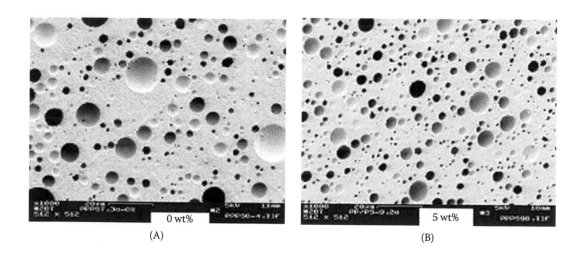

FIGURE 4.15

SEM, smoothed cuts, PS phase etched with tetrahydrofuran (THF), bar = 20 μm. Influence of concentration, tacticity, and side arm length on compatibilizing efficiency of poly(propene-g-styrene) graft copolymers in PP/PS = 2/1 (by wt.) blends. (A), (B), (C), and (D): influence of the compatibilizer content on the PP/PS morphology. Polymer blends PP/PS = 2/1 (by wt.) with varied isotactic PP-PS copolymer (7.4 wt% PS side arms with Mn = 18.000 g/mol) content (SEM, smoothed cuts, PS etched with THF, scale bar = 20 μm). (E), (F), (G), (H), and (I): influence of the copolymer structure on the blend morphology. Polymer blends PP/PS = 2/1 (by wt.) with 5 wt% isotactic copolymer (E), (F), and (G) or atactic copolymer (H and I). (From U. Schulze, T. Fonagy, H. Komber, G. Pompe, J. Pionteck, and B. Ivan, *Macromolecules* 36, 4719–4726, 2003. With permission.)

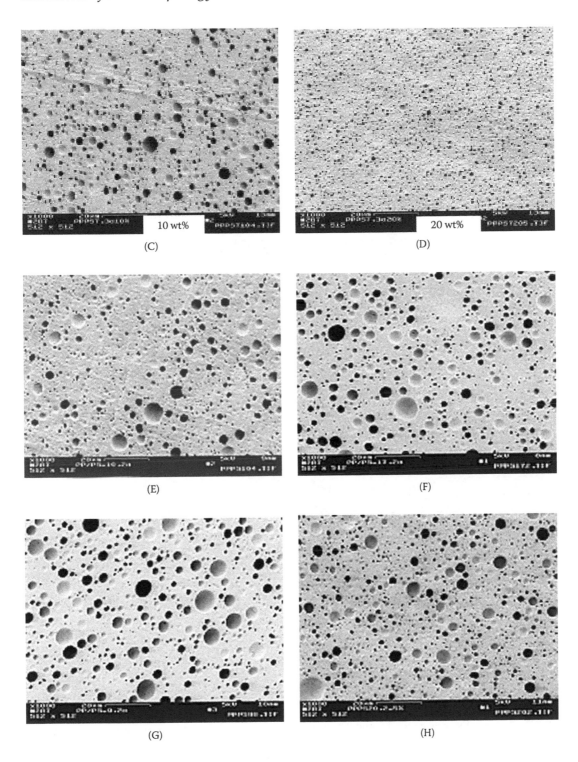

10 wt%
(C)

20 wt%
(D)

(E)

(F)

(G)

(H)

FIGURE 4.15
(*Continued*)

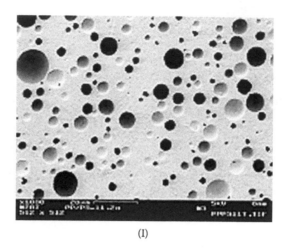

(I)

FIGURE 4.15
(*Continued*)

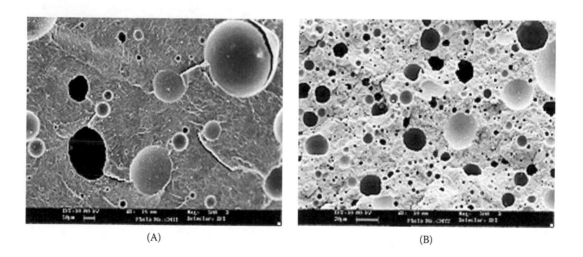

(A)   (B)

FIGURE 4.16
Influence of reactive group concentration on morphology of PP/PS = 2/1 blends. PP is nonreactive, PP-co-Ox3 is reactive to PS-COOH, the compositions and reactive group Ox3/COOH mol ratios are given below the SEM images (cryofractures, PS-phase etched with THF, frame size 233 × 187 m). (A): PP/PS-COOH (2/1 by wt., non-reactive); (B): PP/PP-co-Ox3/PS-COOH (60/6.7/33.3) reactive group ratio 0.5; (C): PP/PP-co-Ox3/PS-COOH (41.7/25/33.3) reactive group ratio 1.8; and (D): PP/PP-co-Ox3/PS-COOH (0.0/66.7/33.3) reactive group ratio 4.8. (From A. Kaya, G. Pompe, U. Schulze, B. Voit, and J. Pionteck, *J. Appl. Polym. Sci.* 86, 2174–2181, 2002. With permission.)

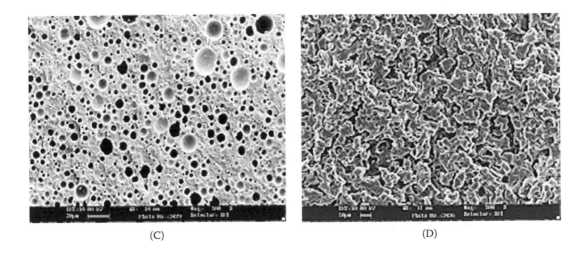

(C)   (D)

FIGURE 4.16
(Continued)

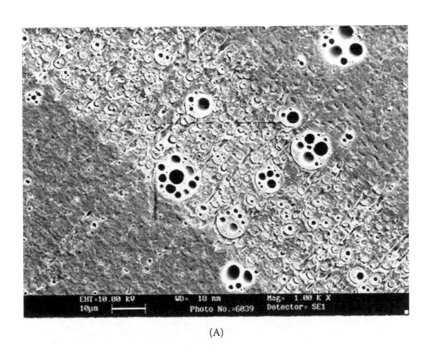

(A)

FIGURE 4.17
Morphology of a ternary system of PP-Ox/PS-Ox (2/1 by wt.) plus hyperbranched polymer (HBP, 5 wt%) obtained by simultaneous mixing for 5 min (A) or 30 min (B). The HBP enriches in the (bright) PS phase resulting in a salami-like phase in phase morphology. (DACA Micro Compounder; SEM, cryocut, HBP etched out with NaOH solution.) (From J. Pionteck, P. Pötschke, N. Proske, H. Zhao, H. Malz, D. Beyerlein, U. Schulze, and B. Voit, *Macromol. Symp.* 198, 209–220, 2003. With permission.)

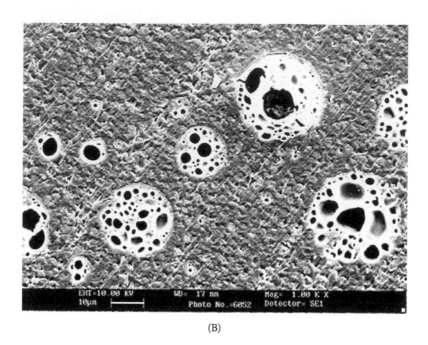

(B)

FIGURE 4.17
(*Continued*)

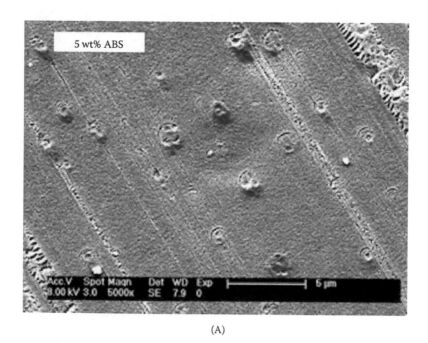

(A)

FIGURE 4.18
SEM photomicrographs of cured epoxy-acrylobutadienestyrene (ABS) blends (scanning electron microscope XL 30 ESEM-FEG [Philips] on cut surfaces, etched with oxygen plasma, and sputtered with gold). (From Y. Müller, L. Häußler, and J. Pionteck, *Macromol. Symp.* 254, 267–273, 2007. With permission.)

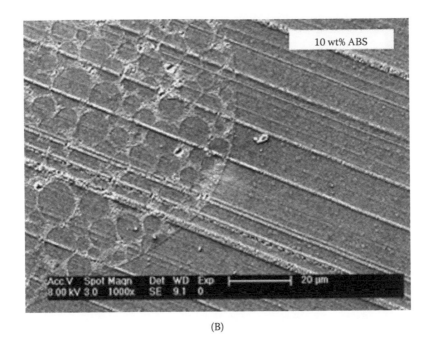

(B)

(C)

FIGURE 4.18
(*Continued*)

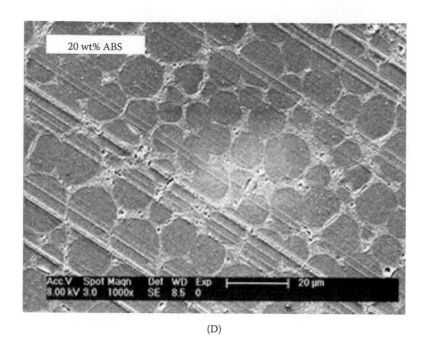

(D)

FIGURE 4.18
(*Continued*)

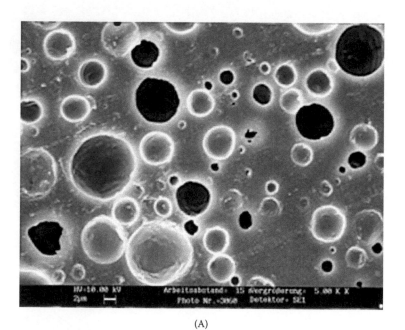

(A)

FIGURE 4.19
Influence of reactive site concentration and viscosity ratio (lambda) on morphology of PP-g-Ox/PS-COOH = 70/30 (wt)-blends (PS was etched out with xylene at room temperature, cryocut). (A): PP-g-Ox/PS, lambda = 0.12, COOH: Ox = 0; (B): PP-g-Ox/PS-COOH 1, lambda = 0.11, COOH: Ox = 0.51; and (C): PP-g-Ox/PS-COOH 2, lambda = 1.03, COOH: Ox = 0.30. (From P. Pötschke, H. Malz, and J. Pionteck, *Macromol. Symp.* 149, 231–236, 2007. With permission.)

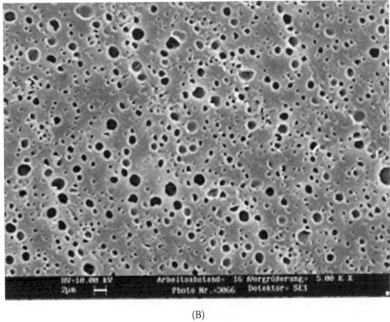

(B)

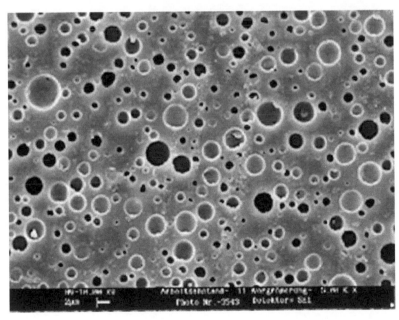

(C)

FIGURE 4.19
(*Continued*)

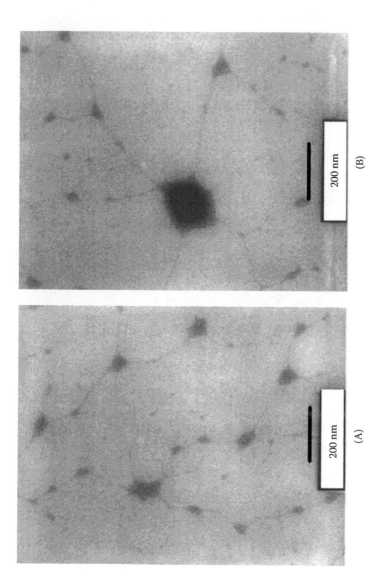

FIGURE 4.20
Transmission electron microscopy results for OsO4 stained samples of UV-cured network based on polypropylene oxide (PPO 2000)/2,4 toluene diisocyanate/2-hydroxyethyl acrylate (TDI/HEA) precursor. The photocured networks of cross-linked polyurethane acrylates based on PPO 2000/2,4 TDI/HEA UV-cured for 150 sec exhibit the phase morphology displayed in (A) and (B). The TEM images clearly show an inhomogeneous structure for the networks, with rather large microgel-like clusters displayed as dark zones on the photomicrographs and connected with each other, forming a sort of "spider's web." (From P. H. Barbeau, J. P. Pascault, and G. Vigier, *J. Polym. Sci.: Part B: Polym. Phys.* 37, 919–937, 1999. With permission.)

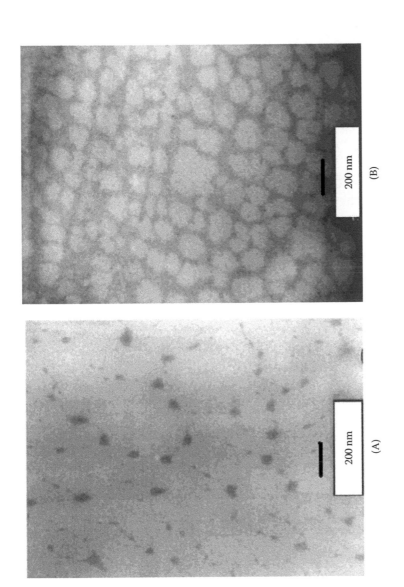

(A)

(B)

FIGURE 4.21

Transmission electron microscopy results for OsO4 stained samples of PPO 2000/isophorone diisocyanate (IPDI)/HEA polymerized for 150 sec, containing (A) 2 wt% and (B) 47 wt% of diacrylated diisocyanate (DAD). The image compares the TEM images obtained for networks based on PPO 2000/IPDI/HEA precursors and containing various amounts of diacrylated diisocyanate HEA/IPDI/HEA molecules. The result is spectacular: the "loose" interconnected structure observed in two dimensions with the network containing 2 wt% of DAD strongly tightens when the DAD concentration increases. In (B), the "alveolar" morphology of the network suggested previously is clearly displayed in the observation plan for the system containing 47 wt% of DAD (in these conditions, 85% of the CAC are then brought by the DAD molecules). The "cells" that are thus delimited correspond undoubtedly to the "polyurethane" phase (macrodiol chains end-capped by diisocyanate units). Moreover, this behavior suggests that the nature of the connecting lines and the large aggregates are close to each other: increasing the DAD concentration leads to a thickening of the walls of the honeycomb. In the case of 2,4 TDI–based networks, the same study could not be done because the HEA/2,4 TDI/HEA segments are crystalline at room temperature (Tm = 68°C) and thus the solubility of the incorporated DAD in the diacrylated prepolymer is very low. (From P. H. Barbeau, J. F. Gerard, B. Magny, J. P. Pascault, and G. Vigier, *J. Polym. Sci.: Part B: Polym. Phys.* 37, 919–937, 1999. With permission.)

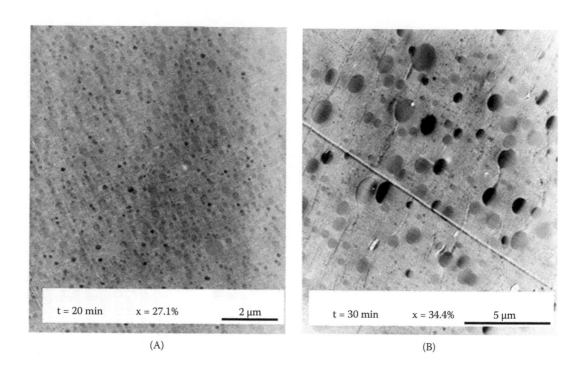

(A)   (B)

FIGURE 4.22

Transmission electron photomicrographs of the PS/4,4′-methylenebis(2,6-diethylaniline)/di-glycidyl ether of bisphenol A (DGEBA/MDEA) (60/40) system at different stages of phase separation induced by polymerization at 177°C in quiescent conditions. (A): 20 sec; (B): 30 sec; (C): 35 sec; (D): 40 sec; (E): 66 sec; and (F): 360 sec. The characteristics of the systems are conversion to cloud point (xcp) ¼ 0:27; 0:56, conversion to gelatin point (xgel), 0:63. These images show the evolution of the morphology observed by TEM of the PS/DGEBA/MDEA (60/40) blend under quiescent conditions at 177°C. Blends containing 60 wt% PS exhibited morphologies consisting of a dispersion of spherical thermoset-rich particles in a thermoplastic matrix. The shape remains spherical during the entire process. The particles grow during the period from 20 to 55 min, and then their growth rate decreases, leading to a final diameter of 2.8 mm. The morphology continues to evolve even when the gel point of the dispersed phase is overtaken. The surface fraction increases, indicating the growth of particles or the appearance of new particles. (From L. Meynie, F. Fenouillot, and J.-P. Pascault, *Polymer* 45, 1867–1877, 2004. With permission.)

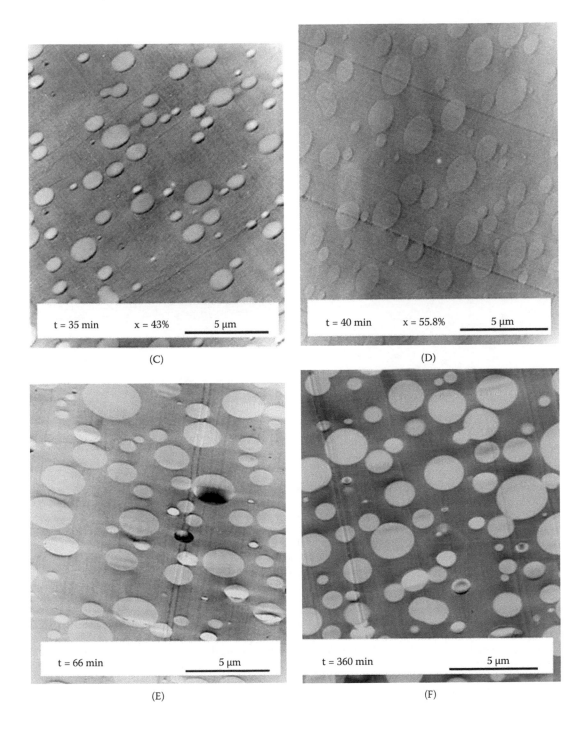

t = 35 min   x = 43%   5 μm

(C)

t = 40 min   x = 55.8%   5 μm

(D)

t = 66 min   5 μm

(E)

t = 360 min   5 μm

(F)

FIGURE 4.22
(*Continued*)

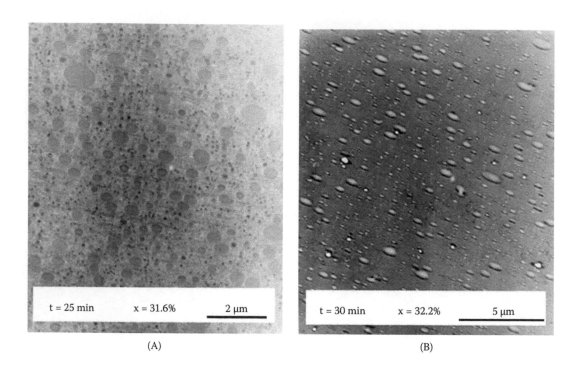

(A)   (B)

FIGURE 4.23

Transmission electron photomicrographs of the PS/DGEBA-MDEA (60/40) system at different stages of phase separation induced by polymerization at 177°C in dynamic conditions. (A): 25 sec; (B): 30 sec; (C): 40 sec; (D): 43 sec; (E): 49 sec; (F): 80 sec. The characteristics of the systems are xcp ¼ 0:27; 0:56, xgel, 0:63. In contrast to Figure 4.29, the TEM image of this figure represents the morphology of the blend PS/DGEBA-MDEA (60/40) under dynamic conditions at 177°C. Quantitative analysis of the morphologies for different reaction times has been reported in reference cited below. (a) During the first 35 min of mixing, the morphology of the blend consists of a dispersion of spherical epoxy-rich particles. The area of the particles does not exceed 2 mm2 until 30 min of shearing. As far as the dispersed particles are viewed as a viscous liquid (before the gel point), the mechanism of coalescence and breakup is possible as in classical thermoplastic/thermoplastic blends. Thus the shape of the particles remains spherical because the process of shape relaxation is possible. Compared to the size of the particles under static conditions (Figure 4.22), those obtained under shear are smaller, indicating that the breakup process occurs and that it is more important than the coalescence process. Around 43 min, the time for epoxy gelation, the particle area distribution broadens significantly, leading to a bimodal particle size distribution consisting of small spherical particles. The first population with small size is composed of newly nucleated particles that have nucleated in a very viscous matrix. We assume that these small particles are formed continuously until the gelation of the dispersed phase rather than formed by the tip-streaming process. These small particles coalesce preferentially and rapidly because the time to drain the film is shorter for small particles. (From L. Meynie, F. Fenouillot, and J.-P. Pascault, *Polymer* 45, 1867–1877, 2004. With permission.)

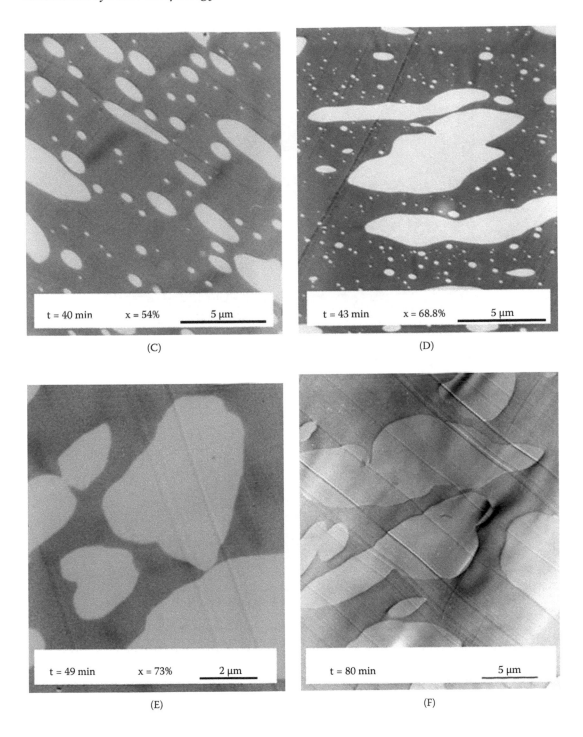

FIGURE 4.23
(*Continued*)

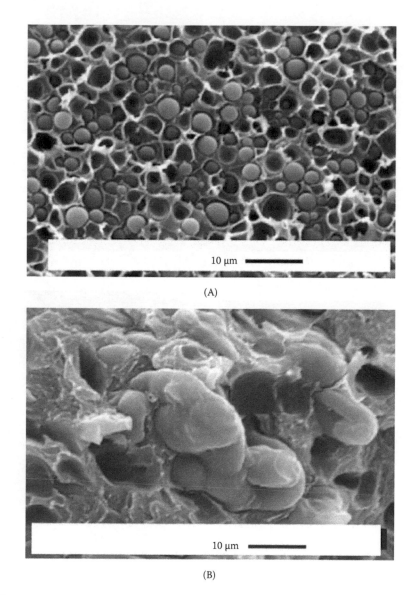

(A)

(B)

FIGURE 4.24
SEM photomicrographs of the final morphology of PS/epoxy-amine 60/40 blend after the polymerization of the epoxy at 177°C. (A): in static conditions, in oil bath. (B): in dynamic conditions, in the mixer. (From L. Meynie, F. Fenouillot, and J.-P. Pascault, *Polymer* 45, 5101–5109, 2004. With permission.)

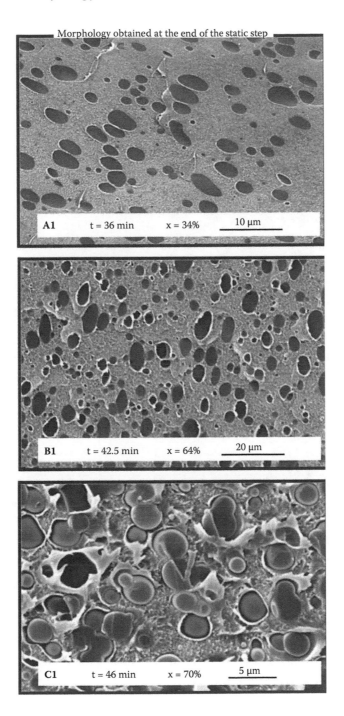

FIGURE 4.25

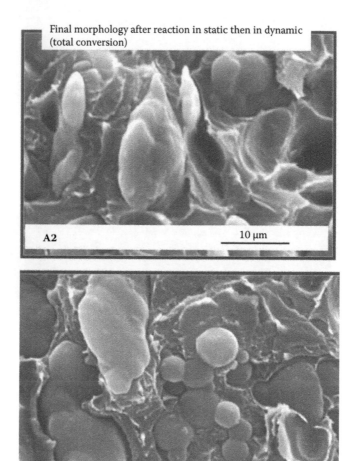

FIGURE 4.25
(*Continued*)

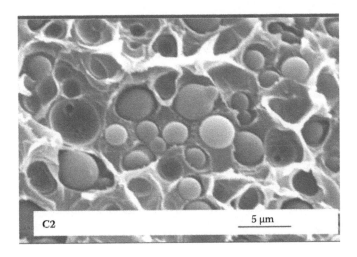

FIGURE 4.25 (*Continued*)

SEM micrographs of the PS/DGEBA/MDEA (60/40) blend system. Static/dynamic coupling (A) before the gelation of the dispersed phase, (B) during the gelation of the dispersed phase, and (C) after the gelation of the dispersed phase. The index 1 indicates the morphologies obtained at the end of the static step. The index 2 indicates the final morphology after reaction in static, then in dynamic; x is the epoxy functions conversion. For (A), (B), and (C) situations, the morphologies obtained in static conditions (A1), (B1), and (C1) (before having restarted the mixer) and the final morphologies (A2), (B2), and (C2) (at the end of the epoxy reaction) are presented in this figure. For the three situations studied, the morphologies obtained just before restarting the rotors of the mixer consists of a dispersion of spherical drops (A1), (B1), and (C1). This was expected, as the epoxy reaction and phase separation proceed in static. The orientation of the droplets in (A1) and (B1) is produced by the action of taking the sample off the chamber because it is difficult to avoid any deformation, especially in the first stages of the polymerization when the polymer is not too viscous. More interesting is the analysis of the final morphologies, after the dynamic stage. (1) The final morphology obtained when the rotors are started just before the gel point is the same as the one observed in purely dynamic conditions (compare Figure 4.24B and [A2]). The fact that the epoxy polymerization has proceeded in the absence of shear for 36 min affects neither the size nor the shape of the final particles, compared to an experiment run totally under shear. Because at 36 min the DGEBA-MDEA–rich phase is still a liquid, when shear is introduced, nothing prevents the agglomeration. (2) The morphology obtained when the rotors are started 3.5 min after the gel point (situation c) is the same as the one observed in purely static conditions. The shape and the size of the particles are the same (compare Figure 4.24A with [C2]). (From L. Meynie, F. Fenouillot, and J.-P. Pascault, *Polymer* 45, 1867–1877, 2004. With permission.)

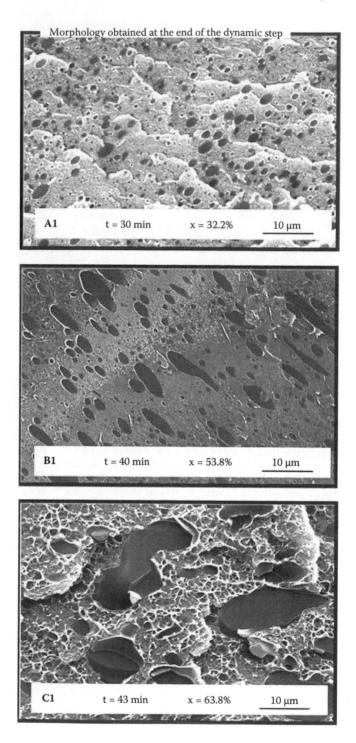

FIGURE 4.26

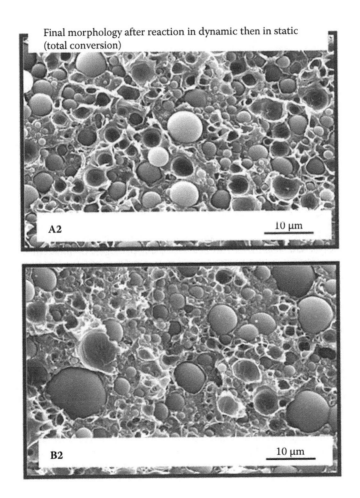

FIGURE 4.26
(*Continued*)

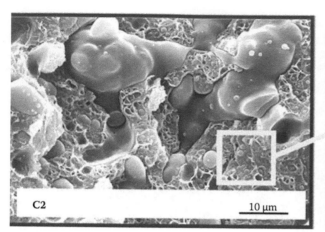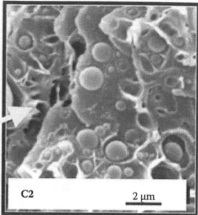

FIGURE 4.26 *(Continued)*

SEM micrographs of the PS/DGEBA/MDEA (60/40) system. Dynamic/static coupling: (A) and (B) before the gelation, (C) during the gelation. The index 1 indicates the morphologies obtained at the end of the dynamic step. The index 2 indicates the final morphology after reaction in dynamic, then in static; x is the epoxy functions conversion. This figure shows the morphologies obtained in dynamic conditions, the corresponding one, and those obtained after the postcure. The dynamic/static experiments are not as information-rich as the static/dynamic; however, we can point out the following observations: (1) For situations (A) and (B), for which the polymerization proceeded under shear for, respectively, 30 and 40 min, the final morphology after having finished the polymerization in static is composed of spherical particles. This was expected, as the rotors are stopped before the gel point of the epoxy-rich phase, so that the droplets are able to undergo relaxation of shape in order to minimize the interfacial energy of the system. However, the size distribution is broader than for the purely static experiment of Figure 4.24A. The explanation is the breakup and coalescence phenomena experienced by the drops under shear, while they are liquid. Hence, smaller particles are produced (breakup) and also larger ones (coalescence). (2) When the reaction is run under shear for 43 min (situation [C]), the coalescence and agglomeration phenomena have already occurred and the large irregular particles, unable to relax, are observed. A bimodal distribution is observed as in the case of a purely dynamic experiment, with very small spheres coexisting with the large particles. When the reaction is continued at zero shear, the bimodal structure is fixed. It is better revealed on the SEM photomicrographs because the interfacial adhesion is poor at the end of the polymerization. The spherical particles have a diameter inferior to 1.5 mm (zoom of [C2]). They must not be confused with the ones observed in Figure 4.25B2 that are bigger (3 mm) because their origin is different. Here, the very small particles are thought to be nucleated during the late phase separation occurring when the viscosity of the medium increases; as a consequence their epoxy conversion is low, whereas those observed in Figure 4.25B2 had been formed in the early stages of the phase separation and their epoxy conversion is high. (3) Finally, when the rotors are stopped at 46 min—that is, 3.5 min after the gel—the very little spherical drops are no longer observed; they have agglomerated. The static postcure does not change the morphology of the blend because this treatment is done after the gelation of most of the dispersed particles. (From L. Meynie, F. Fenouillot, and J.-P. Pascault, *Polymer* 45, 1867–1877, 2004. With permission.)

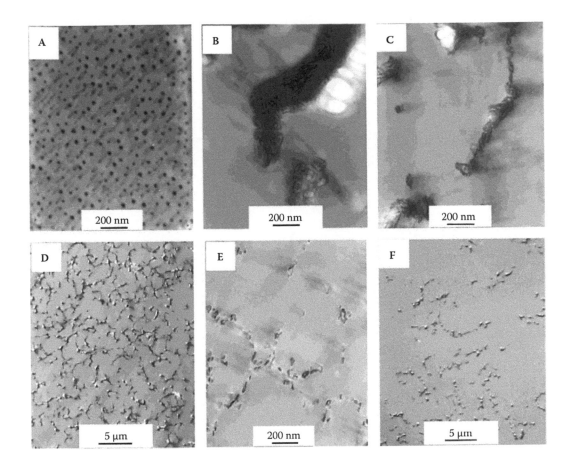

FIGURE 4.27

Typical morphologies of several epoxy thermosets, blended with the nonreactive or slowly reactive terpolymers polystyrene-*block*-polybutadiene-*block*-poly[(methyl methacrylate)-*stat*-(*tert*-butyl methacrylate)] (SB(MT)) or polystyrene-*block*-polybutadiene-*block*-poly[(methyl methacrylate) (SBM). TEM photomicrographs of (A) DGEBA/MCDEA + 10% SB(MT)-77, (B) DGEBA/DDS + 10% SB(MT)-77, (C) DGEBA/MDA + 10% SB(MT)-77, (D) DGEBA/2-PI + 10% SB(MT)-77, (E) DGEBA/MTHPA + 10% SBM-0.03, and (F) DGEBA/DICY + 3% SBM-0.03. Polystyrene-block-polybutadiene-block-poly[(methyl methacrylate)-stat-(methacrylic acid)]. [SB(MA)] block copolymers incorporating acid-reactive functionalities in the last block have been synthesized and studied as modifiers for epoxy thermosets based on the DGEBA. Many techniques TEM have been used to demonstrate the effectiveness of the reaction-induced modification compared to that with the nonreactive or slowly reacting polystyrene-block-polybutadiene-block-poly[(methyl methacrylate)-stat-(tert-butyl methacrylate)] SB(MT) triblock copolymer. Morphological characteristics revealed by TEM indicate that SB(MT) and SB(MA) are both miscible with the epoxy prepolymer. DGEBA/DDS (4,4′-diaminodiphenyl sulfone) as an epoxy-thermoset model system in the presence of poly[(methyl methacrylate)-*stat*-(methacrylic acid)] (HT121) or the block copolymers. The cure of the thermoset/block copolymer system has been explored using six different curing processes, including 2-phenylimidazole (2-PI), alone or in the presence of methyltetrahydrophthalic anhydride (MTHPA) as comonomer; accelerated dicyandiamide (DICY); and three different diamines as comonomers without accelerator: MCDEA, 4,4′-methylenedianiline (MDA), and DDS. The use of reactive block copolymers instead of nonreactive ones permits a better control of morphology. (From V. Rebizant, A.-S. Venet, F. Tournilhac, E. Girard-Reydet, C. Navarro, J.-P. Pascault, and L. Leibler, *Macromolecules* 37, 8017–8027, 2004. With permission.)

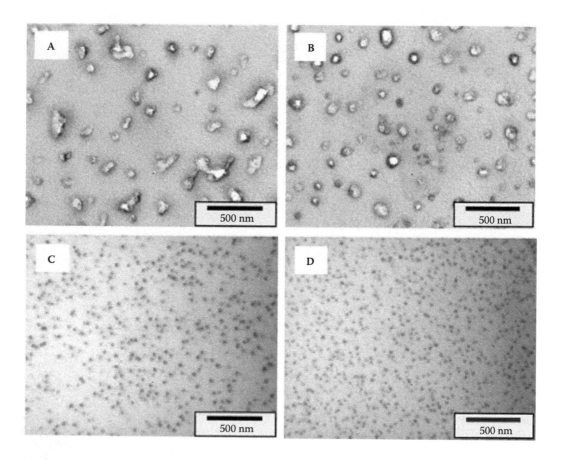

FIGURE 4.28
TEM images of DGEBA/DDS blends modified by 5 wt% of diblock and triblock copolymers. Samples are cured at 135°C + 6 h at 220°C and stained 5 min at room temperature with RuO4 vapors. (A): P[BA-b-(MMA-co-DMA)] 9% DMA; (B): P[(MMA-co-DMA)-b-BA-b-(MMA-co-DMA)] 10% DMA; (C): P[BAb-(MMA-co-DMA)] 29% DMA; and (D): P[(MMA-co-DMA)-b-BA-b-(MMA-co-DMA)] 25% DMA. DGEBA = diglycidyl ether of bisphenol; the hardener (4,4′-diaminodiphenyl sulfone, DDS); MMA = the random copolymers of methyl methacrylate; DMA-dimethylacrylamide. (From S. Maiez-Tribut, J.-P. Pascault, E. R. Soule, J. Borrajo, and R. J. J. Williams, *Macromolecules* 40, 1268–1273, 2007. With permission.)

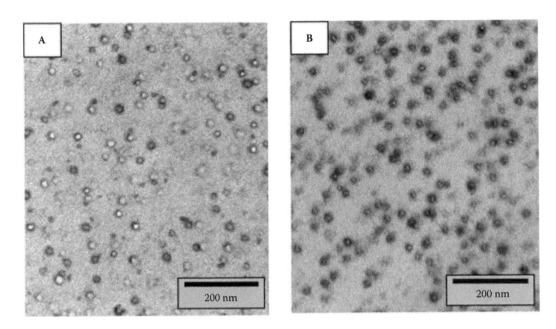

FIGURE 4.29
TEM photomicrographs of fully cured DGEBA/DDS blends with 10 wt% of P[(MMA-co-DMA)-b-BA-b-(MMA-co-DMA)]. Black copolymer (BCP) composition: 33 mol% of BA units, molecular weight of polybutyl acrylate (Mn(PBA)) 20 kg/mol, 25 mol% of DMA units in the random block, and Mn of the triblock 50 kg/mol. Stained with (A) acid phosphotungstic + benzyl alcohol solution; and (B) RuO4 vapors. (From S. Maiez-Tribut, J.-P. Pascault, E. R. Soule, J. Borrajo, and R. J. J. Williams, *Macromolecules* 40, 1268–1273, 2007. With permission.)

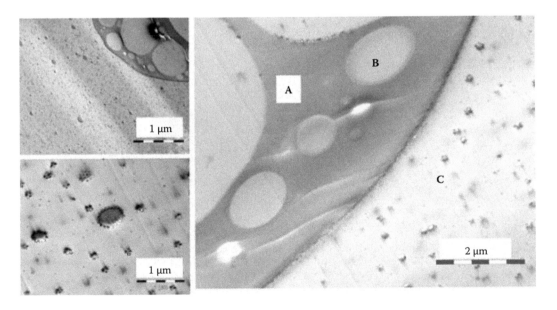

FIGURE 4.30
TEM photomicrographs of fully cured DGEBA/MCDEA blends with 10 wt% of polyphenylene ether (PPE) and modified by 5 pcr of purified poly(styrene-b-butadiene-b-methyl methacrylate) SBM triblock, S2026B26M54. (A) PPE-rich droplet 10 mm; (B) epoxy-rich substructure; and (C) epoxy-rich matrix with dispersed SBM micelles (OsO4 staining). (From Th. Fine and J.-P. Pascault, *Macromol. Symp.* 245–246, 375–385, 2006. With permission.)

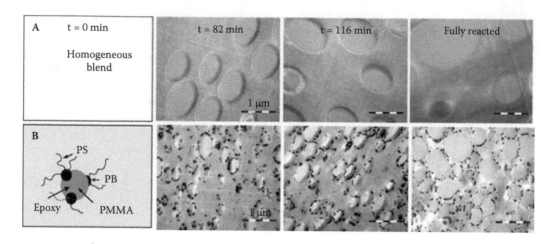

FIGURE 4.31
TEM photomicrographs of DGEBA/MCDEA blends cured at T¼160 8C and observed at different reaction times with (A) 49 wt% of PPE and (B) modified by 10 pcr of purified SBM triblock, S20 26B26M54 (OsO4 staining). (From Th. Fine and J.-P. Pascault, *Macromol. Symp.* 245–246, 375–385, 2006. With permission.)

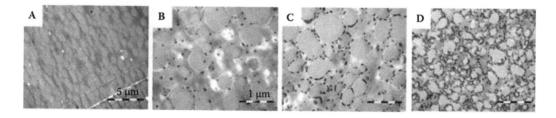

FIGURE 4.32
TEM photomicrographs of (A): fully cured DGEBA/MCDEA blends with 49 wt% of PPE and modified by: (B) 2 pcr, (C) 10 pcr, and (D) 50 pcr of as-received SBM triblock, S20 26B26M54-SB29-S14 (OsO4 staining). (From Th. Fine and J.-P. Pascault, *Macromol. Symp.* 245–246, 375–385, 2006. With permission.)

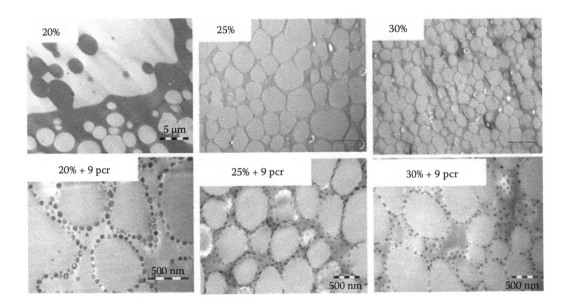

FIGURE 4.33
TEM photomicrographs of fully cured DGEBA/MCDEA blends with different wt% of PPE in the range of phase inversion and modified by 9 pcr of as-received SBM triblock, S20 26B26M54-SB29-S14 (OsO4 staining). (From Th. Fine and J.-P. Pascault, *Macromol. Symp.* 245–246, 375–385, 2006. With permission.)

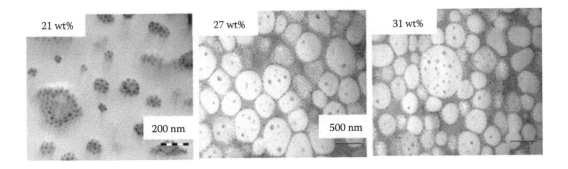

FIGURE 4.34
TEM photomicrographs of fully cured DGEBA/MCDEA blends with different wt% of PPE, in the range of phase inversion and modified by 10 pcr of as-received SBM triblock, S13 16B15M72-SB17-S3 (OsO4 staining). (From Th. Fine and J.-P. Pascault, *Macromol. Symp.* 245–246, 375–385, 2006. With permission.)

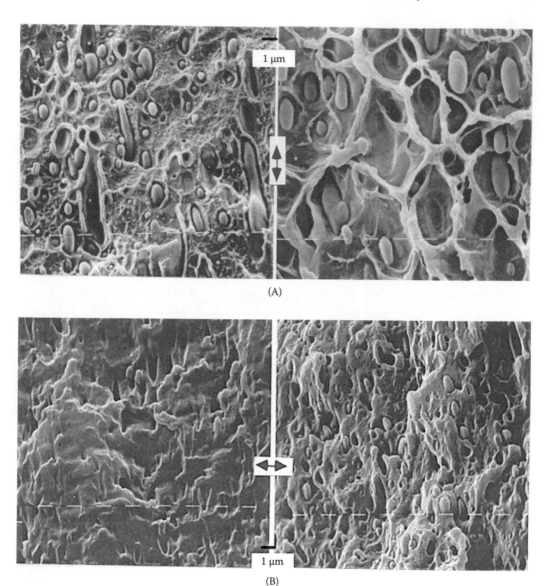

(A)

(B)

FIGURE 4.35

Visualization of the effect of phase morphology and adhesion induced by the hydrogenated polybutadiene block PMMA compatibilizers (molecular characteristics given are total M_n and wt% of PMMA) on the decohesion at the interface in LDPE/PMMA blends. (A): SEM images of cryofractured surfaces of 80/20 LDPE/PMMA blends with 5 wt% copolymer (95k, 33, 4): (left) from a 200% strained tensile specimen and (right) from a completely strained (broken) tensile specimen. (B): SEM images of cryofractured surfaces of 80/20 LDPE/PMMA blends with 5 wt% copolymer (103k, 52): (left) from a 50% strained tensile specimen and (right) from a completely strained (broken) tensile specimen. (C): SEM images of cryofractured surfaces of 80/20 LDPE/PMMA blends with 5 wt% copolymer (95.5k, 71): (top) from a 50% strained tensile specimen and (bottom) from a completely strained (broken) tensile specimen. (From C. Harrats, T. Benabdallah, G. Groeninckx, and R. Jérôme, *J. Polym. Sci.: Part B: Polym. Phys.* 43, 22–34, 2005. With permission.)

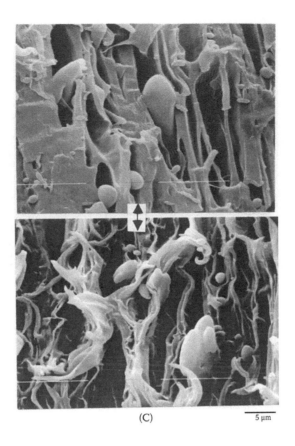

(C)   5 µm

FIGURE 4.35
(*Continued*)

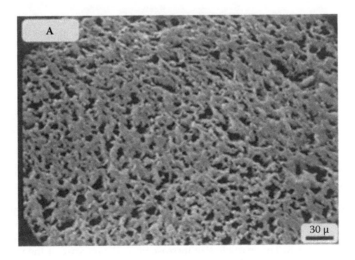

FIGURE 4.36

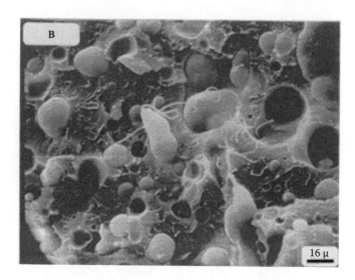

FIGURE 4.36 (*Continued*)

SEM photomicrographs of a 20/80 LDPE/PS blend. (A): the continuous phase remaining after the extraction of PS (roll-milled blend), and (B): the room-temperature fracture surface of the blend (roll-milled and compression-molded for 3 min at 200°C). The melt-blending of 20 wt% LDPE (Shell 33) with 80 wt% PS (polystyrol 158 K) with a two-roll mill at 200°C for 3 min results in a cocontinuous two-phase morphology. However, this morphology is unstable and progressively evolves into a droplet/matrix phase morphology as a result of the breakup, retraction, and coalescence of the elongated structures when the sample is heated at 180°C in the absence of shear as confirmed in the photomicrographs. (From C. Harrats, R. Fayt, R. Jérôme, and S. Blacher, *J. Polym. Sci.: Part B: Polym. Phys.* 41, 202–216, 2003. With permission.)

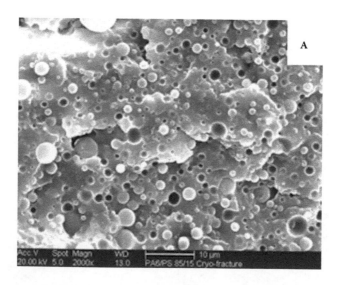

FIGURE 4.37

SEM photomicrographs showing the phase morphology of unmodified melt-blended 85PA6/15PS (A, C) and 70PA6/30PS (B, D) blends: A, B = cryofractured surface; C, D = cryosmoothed, PS phase extracted using chloroform. Average particle sizes of 1.8 and 2 μm are obtained in the absence of compatibilizer in 85PA6/15PS and 70PA6/30PS blends, respectively. This particle size increase upon the increase of the minor phase content is expected, as the process of particle–particle coalescence is favored by increasing the minor phase concentration. (From T. S. Omonov, C. Harrats, and G. Groeninckx, *Polymer* 46, 12322–12336, 2005. With permission.)

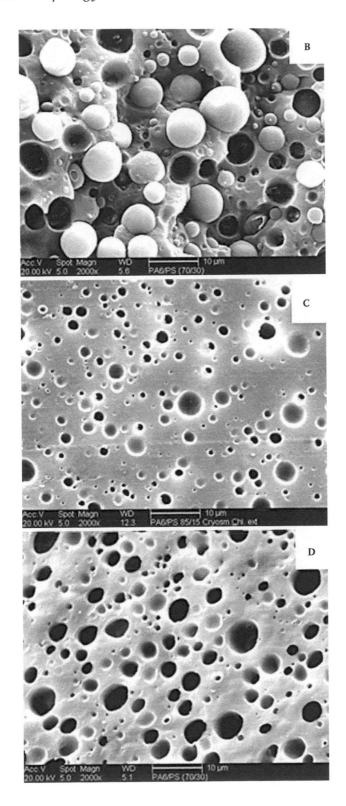

FIGURE 4.37
(*Continued*)

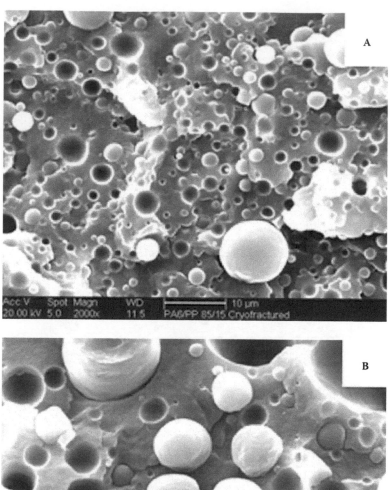

FIGURE 4.38
SEM photomicrographs showing the phase morphology of melt-blended 85PA6/15PP (A, C), and 70PA6/30PP (B, D) blends: (A, B) cryofractured surface, (C, D) cryosmoothed surfaces. The former blend exhibits an average particle size of 2.3 mm, whereas, as expected, coarser PP particles having an average diameter of 3.2 mm are obtained in the latter blend. For the same reasons of favored coalescence, a broader particle size distribution is observed in the blends containing 30 wt% PP compared to blends containing only 15 wt%. (From T. S. Omonov, C. Harrats, and G. Groeninckx, *Polymer* 46, 12322–12336, 2005. With permission.)

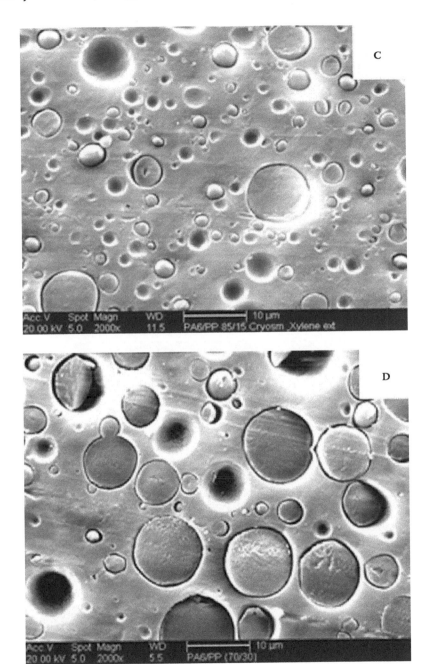

FIGURE 4.38
(*Continued*)

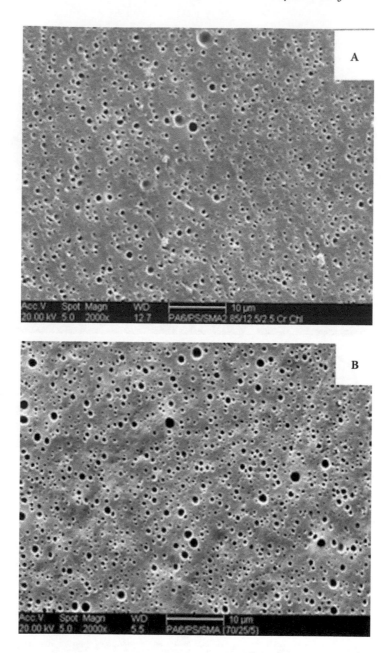

FIGURE 4.39
SEM photomicrographs showing the phase morphology of 85PA6/(12.5/2.5)(PS/SMA2) blends (A), and 70PA6/
(25/5)(PS/SMA2) blends (B). The PS phase was extracted using chloroform. Addition of 2.5 wt% of the reactive precursor SMA2 containing 2 wt% of maleic anhydride to the PA6/PS binary blend reduced significantly the average dimension of the PS dispersed phase at both blend compositions. The sizes of the PS particles are reduced by factors of 3 and 5 in 85/15 and 70/30 compatibilized blends, respectively. The particle size distribution gets narrower than that in uncompatibilized blends. That is a clear indication of the *in situ* generation of the grafted PA6-g-PS copolymer during melt-blending as a result of the amine–maleic anhydride imidation reaction. (From T. S. Omonov, C. Harrats, and G. Groeninckx, *Polymer* 46, 12322–12336, 2005. With permission.)

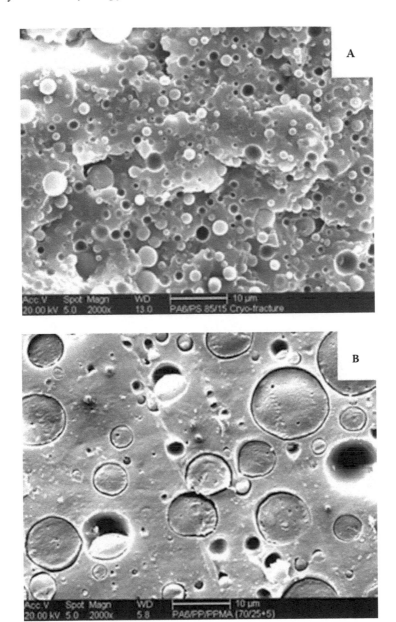

FIGURE 4.40

SEM photomicrographs showing the effect of copolymer on the phase morphology of PA6/PP blends. Cryosmoothed surfaces of 85PA6/(12.5/2.5)(PP/PP–MA2) (A) and 70PA6 /(25/5)(PP/PP-MA2) (B) blends. In contrast, the PP-MA2, which also contains 2 wt% of maleic anhydride groups, does not produce a comparable (as SMA2 in PA6/PS, Figure 4.39) extent of size reduction of the PP particles in the PA6/PP compatibilized blend. Only a slight phase size decrease is recorded in the 85/15 blend (2 μm compared to 2.3 μm in the uncompatibilized blend). In fact, at equivalent maleic anhydride content and total amount of precursor, SMA2 produces dispersed particles that are much smaller in the PA6/PS blend than those formed from the PP-MA2 compatibilizer precursor in the PA6/PP blend. Probably the SMA2 copolymer more easily diffuses and reacts at the interface with the PA6 than does PP-MA2. (From T. S. Omonov, C. Harrats, and G. Groeninckx, *Polymer* 46, 12322–12336, 2005. With permission.)

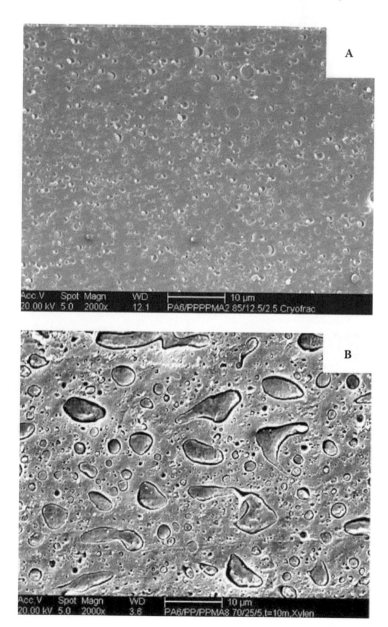

FIGURE 4.41
SEM photomicrographs showing the phase morphology of cryosmoothed surfaces of 85PA6/(12.5/2.5)(PP/PP-MA8) (A), and 70PA6/(25/5)(PP/PP-MA8) (B) blends. When the MA content was increased to 8 wt% (PP-MA8), the compatibilizing effect is tremendous compared to PP-MA2 but still remains less efficient than SMA2 in PA6/PS. Note that the particle size distribution induced by PP-MA8 exhibits a bimodal pattern due to the formation of larger and nonspherical particles as visible in the SEM micrograph of (B). (From T. S. Omonov, C. Harrats, and G. Groeninckx, *Polymer* 46, 12322–12336, 2005. With permission.)

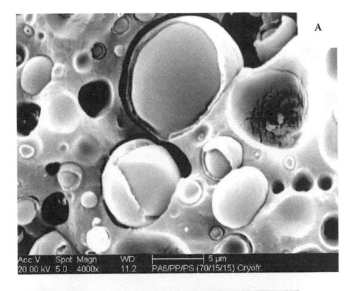

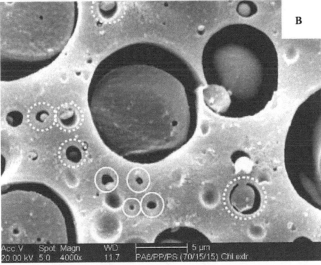

FIGURE 4.42

SEM photomicrographs showing the phase morphology of ternary 70PA6/15PP/15PS blends: (A) cryofractured; (B) chloroform-extracted surfaces. These images reveal the existence of encapsulated droplets dispersed in PA6 matrix. The shells surrounding the core particles are broken upon the cryofracturing of the sample. From this image, it is not possible to determine which phase is where. (B) is a SEM micrograph of a cryosmoothed surface of the same blend where the polystyrene phase was etched using chloroform. This test shows that the shell of the encapsulated structures consists of polystyrene and the core is the polypropylene phase. The average shell thickness is less than 1 μm but greater than 0.2 μm. Close observation of the SEM photomicrograph reveals that isolated polystyrene particles (see white circles) as well as partly encapsulated polypropylene particles (see dashed circles) were also formed during the blend compounding process. (From T. S. Omonov, C. Harrats, and G. Groeninckx, *Polymer* 46, 12322–12336, 2005. With permission.)

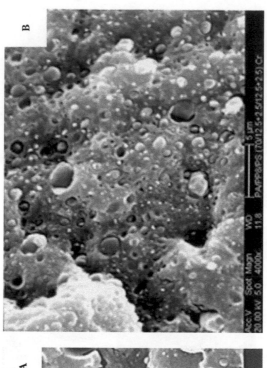

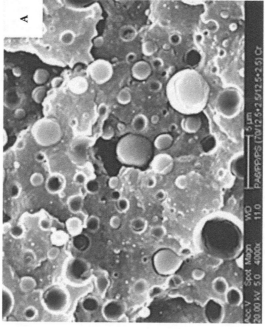

FIGURE 4.43

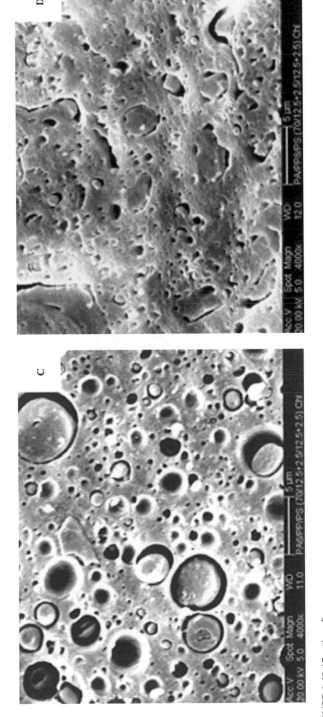

FIGURE 4.43 *(Continued)*

SEM photomicrographs showing phase morphology of polyamide 6/polypropylene/polystyrene modified with styrene maleic anhydride grafted polystyrene. The indices next to MA indicate the percentage of maleic anhydride in the copolymer. 70PA6/(12.5/2.5) (PP/PP-MA2)/(12.5/2.5)(PS/SMA2) (A, C) and 70PA6/(12.5/2.5)(PP/PP-MA8)/(12.5/2.5)(PS/SMA2) (B, D) blends: (A, B) cryofractured, (C, D) cryosmoothed surfaces. Addition of 2.5 wt% of each of SMA2 and PP-MA reactive precursors for the compatibilization of PA6/PS and PA6/PP, respectively, modified significantly the developed phase morphology. The selective etching of the polystyrene phase from the cryosmoothed surfaces using chloroform shows the remaining PP phase in PA6 matrix (C). This image reveals a mixed situation of phase morphology—that is, some of the PP particles are encapsulated by the PS phase and a large number of isolated PS particles are visible as dark holes. Note also that the encapsulated PP particles exhibit a broad particle size distribution. Rough volume estimation shows that about 70% of the polystyrene phase forms its own domains as isolated particles dispersed in the PA6 matrix. By changing the PP-MA2 by a more functionalized PP-MA containing 8 wt% of reactive maleic anhydride groups (PP-MA8), the phase morphology exhibits a significant change. The particle size of the dispersed phase is much smaller than that of the blend modified with PP-MA2 (compare [C] and [D]). The superiority of compatibilization efficiency induced by PP-MA8 over PP-MA1 has been demonstrated previously in a binary blend of PP with polycyclohexylmethacrylate (PCHMA) containing amine functionalized polystyrene PS-NH2. (C) shows that the encapsulation of PP particles by the PS phase is partial. Indeed, small particles of polystyrene are located at the interface between the PP particles and the PA6 matrix. (From T. S. Omonov, C. Harrats, and G. Groeninckx, *Polymer* 46, 12322–12336, 2005. With permission.)

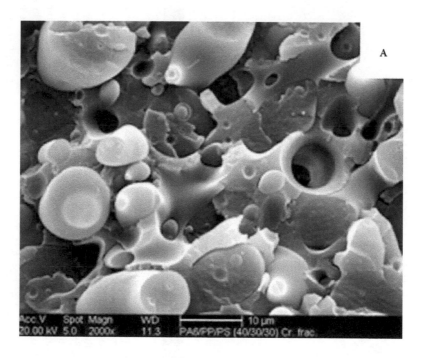

FIGURE 4.44
SEM photomicrographs showing the phase morphology of uncompatibilized 40PA6/30PP/30PS blend. (A): cryofractured; (B): cryosmoothed and chloroform etched; and (C): cryosmoothed and formic acid–etched surfaces. The cryofracture surfaces show complex phase morphology where it is not possible to identify which phase is located where. (B) shows the cryosmoothed surface of the same blend from which the polystyrene phase was selectively extracted using chloroform. This solvent treatment revealed that the PS phase constitutes a continuous phase as indicated by the hollow continuous space in between the other two phases. A small percentage of the PS phase is also dispersed in the other phases. This image does not allow the identification of phases and locations of the subinclusions of the PS phase. The chloroform-extracted sample has been further treated with formic acid, a selective solvent for the PA6 phase. Unfortunately, the remaining PP-based structure was not self-supporting and collapsed surfaces were obtained that do not allow characterizing the state of the phases. The SEM micrograph of (C) is obtained from a sample where only PA6 was selectively extracted. It reveals that the PA6 is in the form of droplets on the observed surface. Does the PA6 phase form a dispersion or is it continuous? The quantitative extraction experiments carried out on the same blends using formic acid confirmed that PA6 is continuous. Indeed, almost all the PA6 phase was extracted from the blend by the formic acid solvent (97%). That means that the PA6 phase is certainly elongated in a network structure in the extrusion direction as we have demonstrated recently in other blend systems. It is thus clear that the three blend components form a three-phase cocontinuous morphology. Furthermore, the extraction data reveal that in the blends of 40PA6/30PP/30PS about 3 wt% of polyamide are dispersed as subinclusions in the PP phase. (From T. S. Omonov, C. Harrats, and G. Groeninckx, *Polymer* 46, 12322–12336, 2005. With permission.)

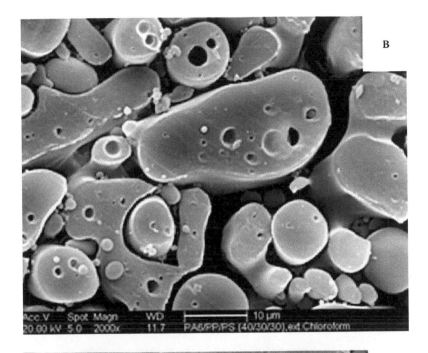

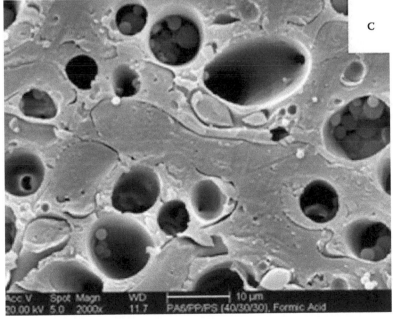

FIGURE 4.44
(*Continued*)

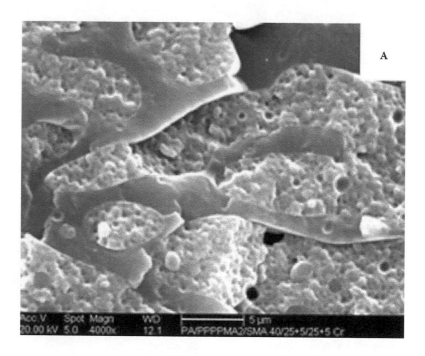

FIGURE 4.45
SEM photomicrographs showing the phase morphology of a 40PA6/(25/5)(PP/PP–MA2)/(25/5)(PS/SMA2)
blend. (A): cryofractured; (B): cryosmoothed and chloroform extracted; and (C): cryosmoothed and formic acid–
extracted surfaces. Upon the substitution of 5 wt% of PP and 5 wt% of PS by their respective compatibilizer
precursors—that is, PP-MA2 or PP-MA8, and SMA2, respectively—the phase morphology has been modified
deeply. The cryofractured surfaces observed do not allow the identification of the phases. (B) shows a SEM
micrograph of the cryosmoothed surfaces of the same blends after extraction of the PS phase. The remaining
two phases, polyamide and polypropylene, seem composed of very dense and interconnected granular particles. Were these particles precipitated as a layer after the removal of the polystyrene phase or do they constitute
the bulk of the remaining polyamide or polypropylene phases? To address this question, a further selective
extraction of a phase is necessary. (C) shows an SEM photomicrograph of the same blend where the PA6 phase
has been removed using formic acid. The observed image reveals very useful information on the state of the
polyamide phase. Indeed, as indicated by the hollow black space on the image, the polyamide phase is mainly
dispersed in the PS phase and also forms a layer in between the PS and PP phases. (From T. S. Omonov, C.
Harrats, and G. Groeninckx, *Polymer* 46, 12322–12336, 2005. With permission.)

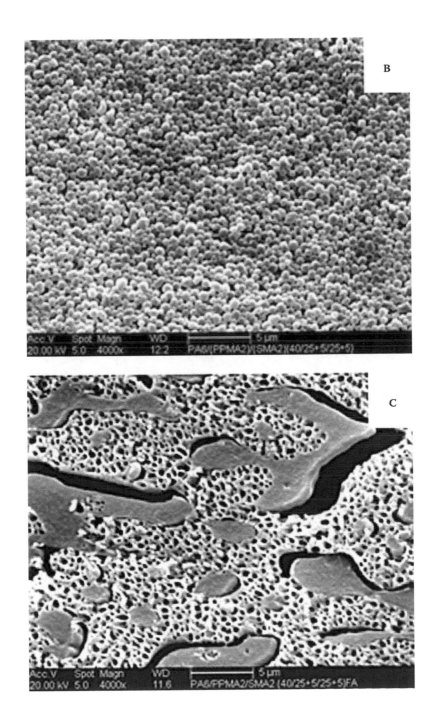

FIGURE 4.45
(*Continued*)

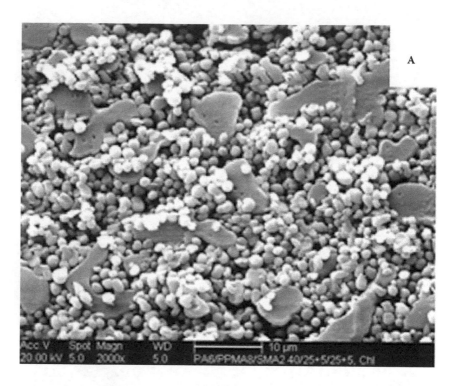

FIGURE 4.46

SEM photomicrographs of 40 PA6/(25/5) (PP/PP–MA8)/(25/5)(PS/SMA2) blends. (A): cryosmoothed and chloroform extracted; (B): formic acid extracted; and (C): PA6 phase extracted with formic acid after PS phase extraction with chloroform. The two remaining phases are clearly visible after the extraction of the PS phase, but no distinction can be made between the PA6 and the PP phases. The SEM micrograph of the blend in which the polyamide phase is extracted is shown in (B). It shows that the PA6 phase is included as granules in either the PP or the PS phase and also entrapped in between them as a thin layer. In order to show the location of PA6, this phase has been extracted with formic acid from a sample from which PS has already been removed. The SEM micrograph of that blend, shown in (C), reveals that the PA6 is preferably included in the PS phase (note that no holes are left on the PP remaining phase). However, a significant difference lies in the size of the granules of the polyamide phase. Indeed, in the blend of 40 PA6/(25/5) (PP/PP-MA2)/(25/5) (PS/SMA2) the PA6 phase has an average particle size of 0.5 μm, whereas in the same composition of the 40 PA6/(25/5) (PP/PP-MA8)/(25/5) (PS/SMA2) blend the particle size is almost three times larger. (From T. S. Omonov, C. Harrats, and G. Groeninckx, *Polymer* 46, 12322–12336, 2005. With permission.)

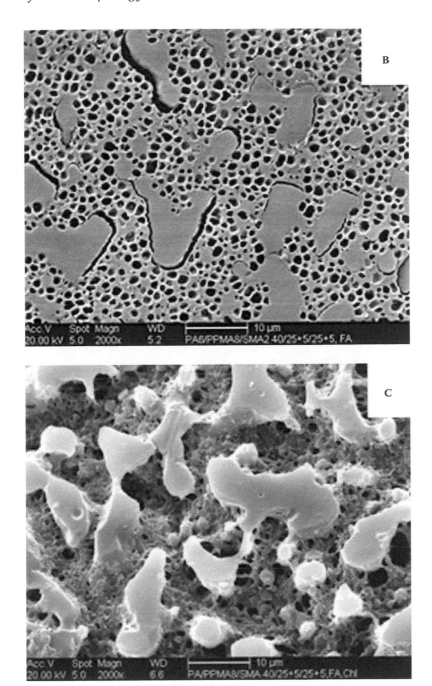

FIGURE 4.46
(*Continued*)

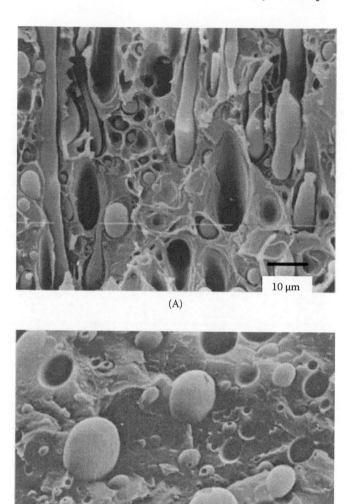

(A)

(B)

FIGURE 4.47

SEM photomicrographs of cryofractured and THF-etched surfaces of 20/80 LDPE/PS added with 10 wt% tri-block copolymer. (A): not annealed; (B): annealed for 150 min at 180°C. In the absence of compatibilizers the LDPE phase exhibited a substantial phase coarsening via a coalescence process. (From C. Harrats, R. Fayt, R. Jérôme, and S. Blacher, *J. Polym. Sci.: Part B: Polym. Phys.* 41, 202–216, 2003. With permission.)

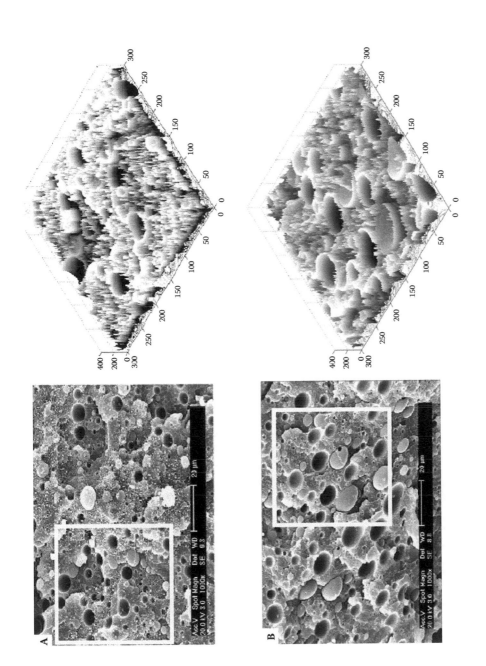

FIGURE 4.48

Phase morphology of melt-blended polypropylene/polystyrene/styrene–butadiene–styrene triblock copolymer blends prepared at different mixing times. (A): 0.5 min; (B): 1.0 min; (C): 2.5 min; and (D): 7.0 min. In these blends, PP is distributed in the matrix in the form of spherites, forming the discrete phase, whereas PS is the continuous phase. The morphology development of polypropylene/polystyrene (PP/PS) blends was studied by means of effective mathematics methods. Time-resolved fracture morphology measurements on PP/PS (20/80) blends compatibilized with styrene–butadiene–styrene (SBS) block copolymer suggested that PP/SBS domains acted as a warehouse supplying compatibilizer (SBS) to the phase boundary in the initial stage of mixing and promoted the formation and development of the transition layer.

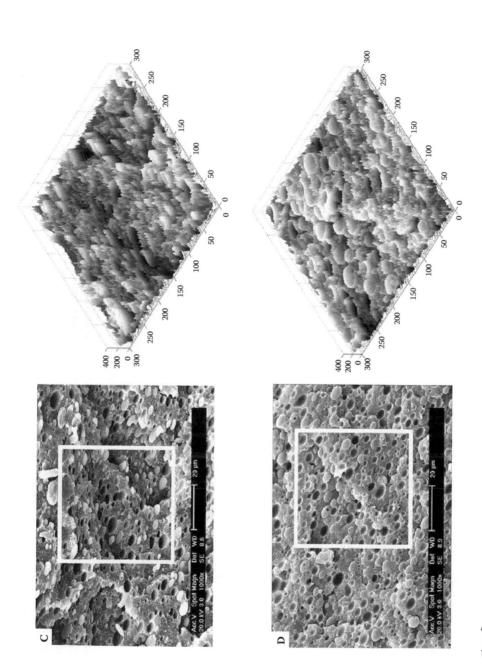

FIGURE 4.48 (*Continued*)

The development of the transition layer led to a more complicated morphology of the fracture surface and strengthened the adhesion between phases, which was quantitatively investigated using Brown fractal dimension DBrown. In the early stage of the mixing (<2.0 min), the mean chord length Km used to describe the domain size decreased; simultaneously, the distribution of K trended to uniform as the mixing proceeded. After 2.0 min, Km fluctuated in a definite range. Further, a normalized distribution of dimensionless domain sizes K/Km was independent of mixing time, indicating that the late stage of phase dispersion can be scaled with a time-depended single length parameter Km. In other words, the morphology development shows a possible dynamic scaling behavior. (From Y.-Y. Li, S.-W. Hu, and J. Sheng, *Eur. Polym. J.* 43, 561–572, 2007. With permission.)

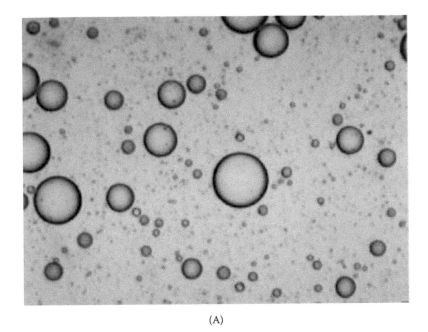

(A)

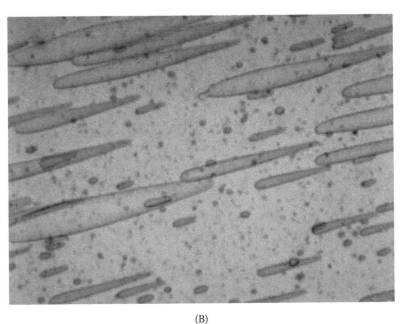

(B)

FIGURE 4.49

OM photomicrographs of a model blend of 50/50 polydimethylsiloxane and polybutadiene (A) before shearing, (B) and (C) during shear up to gamma = 180, and after cessation of shearing by (D) 8 sec, (E) 25 sec, and (F) 89 sec. In (A) the initial microstructure consists of relatively large spherical droplets with a broad distribution of sizes. In (B) and (C) the dispersion is rapidly sheared, and the droplets stretch in the direction of shearing. When the deformation is stopped, the long, thread-like drops initially retract to shorter lengths and then slowly break up (D through F). The final microstructure in (F) has smaller droplets than the original structure, with a narrower size distribution. (From M. Iza and M. Bousmina, *J. Rheol.* 44, 1363–84, 2000. With permission.)

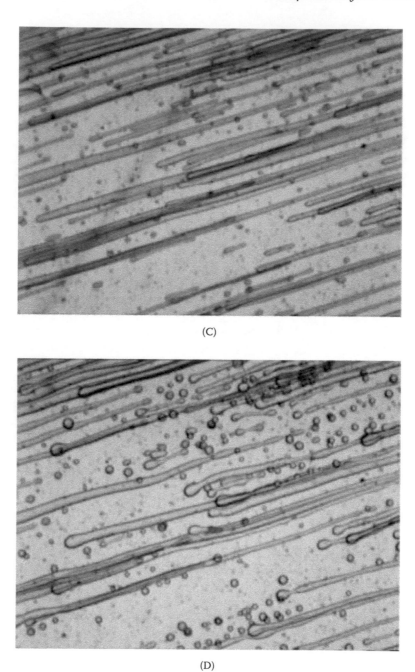

(C)

(D)

FIGURE 4.49
(*Continued*)

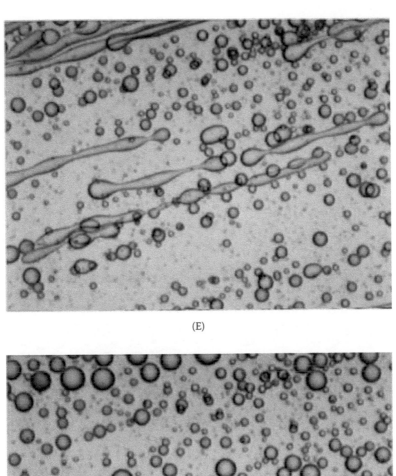

(E)

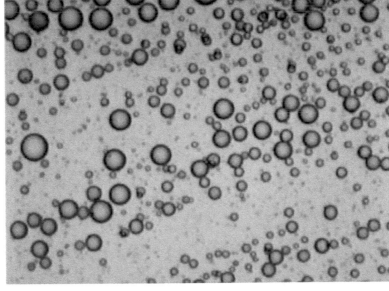

(F)

FIGURE 4.49
(*Continued*)

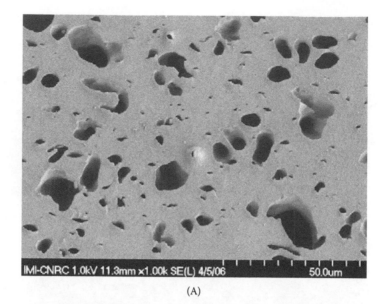

(A)

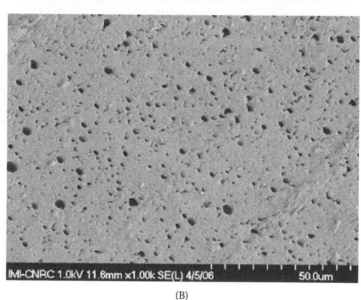

(B)

FIGURE 4.50

SEM photomicrographs of melt-blended thermoplastic starch (TPS) and polylactide (PLA). The microtoming of the surface was carried out at room temperature using a diamond knife, and the surfaces were subsequently treated with hydrochloric acid (HCl, 1 N) for 3 h to selectively dissolve the TPS phase. The micrographs presented in the figure were obtained with the one-step process in which the MA and the peroxide initiator are introduced with the PLA at midextruder through a single-screw extruder side feed. (A) and (B): blends contains 27% TPS; (C) and (D): 43%; (E) and (F): 60%. Blends in (A), (C), and (E) are not compatibilized, whereas the blends in (B), (D), and (E) are modified by free-radical grafting of maleic anhydride (MA) onto the PLA and then by reacting the modified PLA with the starch macromolecules. The effect of compatibilization is very clear on the images. (From M. A. Huneault and H. Li, *Polymer* 48, 270–280, 2007. With permission.)

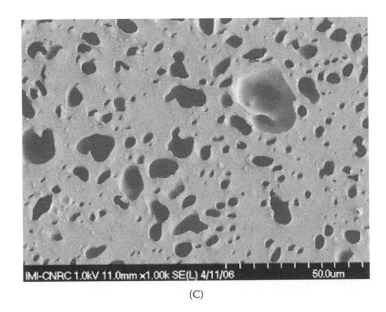

(C)

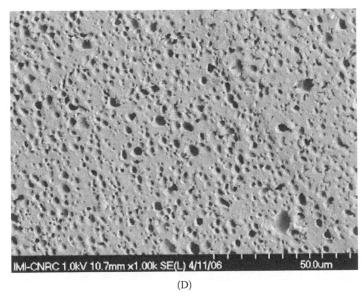

(D)

FIGURE 4.50
(*Continued*)

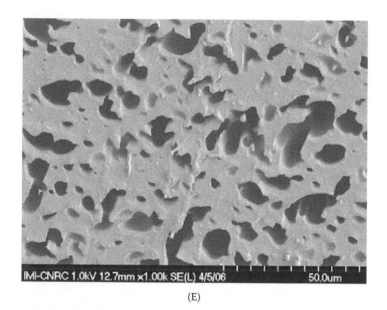

(E)

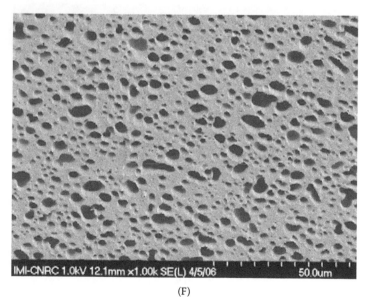

(F)

FIGURE 4.50
(*Continued*)

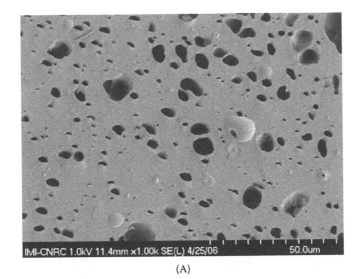

(A)

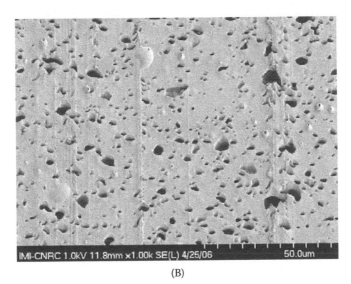

(B)

FIGURE 4.51

SEM photomicrographs for the two-step 27% TPS/PLA blends. The two-step variant of this process consists in first producing the grafted PLA using a dedicated twin-screw process and then substituting in part or full the virgin PLA by the modified PLA in the TPS/PLA blending step. In (A), (B), and (C) 20% of the pure PLA has been replaced by PLA grafted using 2 wt% MA and (A) 0.10, (B) 0.25, and (C) 0.50 wt% L101 (peroxide). In (D) PLA has been fully substituted with the same grafted PLA as in (B). In all cases, the TPS phase size is reduced compared to the unmodified control presented in Figure 4.50A but is not as finely dispersed as the blend prepared using the one-step compatibilization scheme (Figure 4.50B). The smallest and most uniform TPS phase size is observed in (B) when using the 0.25 wt% peroxide level. The reason why higher peroxide level and thus higher grafting degree may not further reduce the TPS phase size is that it comes at the expense of increased PLA chain scission and viscosity reduction. The decreased ability to deform and disperse the TPS phase and the reduction in chain entanglement density between PLA and PLA-g-MA may therefore outweigh the benefit of a slightly higher grafting level. (From M. A. Huneault and H. Li, *Polymer* 48, 270–280, 2007. With permission.)

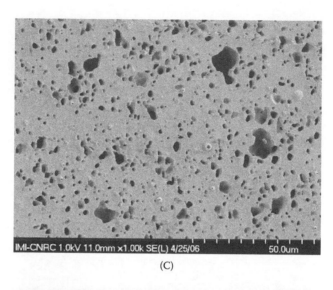

(C)

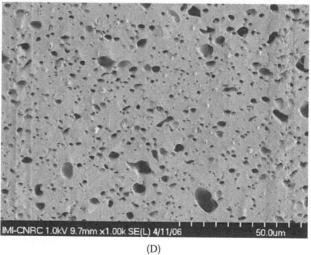

(D)

FIGURE 4.51
(*Continued*)

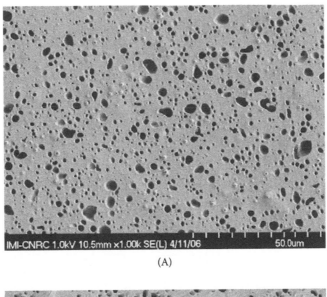

(A)

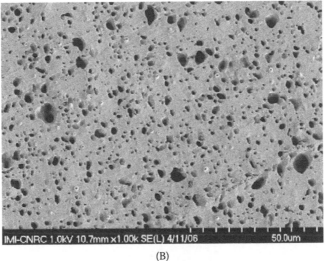

(B)

FIGURE 4.52

SEM photomicrographs of TPS/PLA blends showing the effect of glycerol content in TPS on the phase morphology of 27% TPS/PLA-g-MA blends produced with the one-step process. The glycerol content in the TPS phase is (A) 30%, (B) 33%, and (C) 39%. The TPS comprising 30 wt% glycerol is approximately two times more viscous than the one comprising 33 or 39 wt% glycerol. This limits the deformability of the TPS phase during the mixing step and thus leads to a larger dispersed phase size. (From M. A. Huneault and H. Li, *Polymer* 48, 270–280, 2007. With permission.)

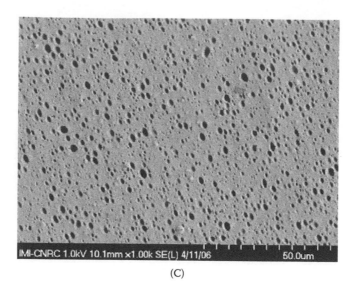

(C)

FIGURE 4.52
(*Continued*)

(A)

FIGURE 4.53

SEM photomicrographs of unmodified and interface-modified 27% TPS/PLA blends where the TPS is based on rice (A and B) or pea (C and D) starch (instead of wheat as in previous figures). The maleic anhydride (MA) and peroxide (L101) concentrations are, respectively, 2 and 0.25 wt% based on total PLA content. Interestingly, the TPS viscosity varies significantly with the starch source, the rice TPS being the most fluid and the pea TPS being the most viscous. This is translated in the presented images into phase morphologies that are much finer for the rice TPS and much coarser for the pea TPS. (From M. A. Huneault and H. Li, *Polymer* 48, 270–280, 2007. With permission.)

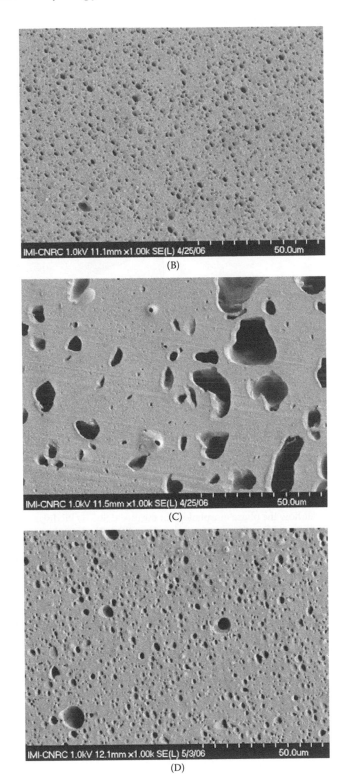

(B)

(C)

(D)

FIGURE 4.53
(*Continued*)

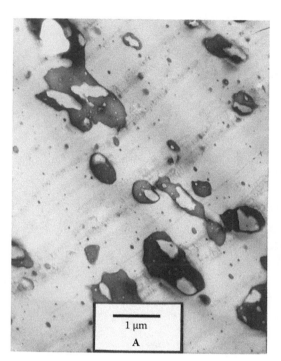 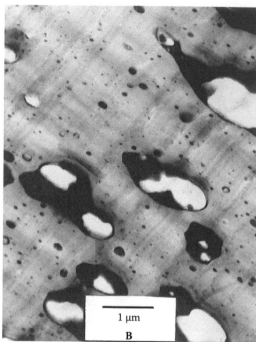

FIGURE 4.54

TEM photomicrographs of polystyrene/styrene-butadiene rubber 1 (SBR-1)/PE blends of different PE contents in the mixed SBR-1/PE dispersed phase: (A) 28%; (B) 38%; (C) 50%; and (D) 64%. SBR has an Mw of 630 k. The series of images presented illustrate how the phase morphology of the mixed SBR-1/PE dispersed phase depends on the PE content. Whatever this content, PE is always encapsulated by SBR-1. The internal morphology of the dispersed phase changes with the SBR/PE weight ratio. When the SBR-1/PE weight ratio is such that PE is dispersed in the SBR-1 phase for the parent binary blends, more than one PE particle is usually observed in the SBR-1 droplets (A, B). When the PE content is increased beyond the composition of phase inversion, much larger and much more irregularly shaped PE subphases are observed (C), which become regular in shape and closely follow the contour line of the SBR-1 phase at PE contents far beyond the phase inversion (D). In this case, some SBR subinclusions are observed within the PE particles. Phase morphology was observed with a Philips CM 100 transmission electron microscope. A Reichert–Jung Ultracut FC 4 microtome cooled at –100°C and equipped with a diamond knife was used to prepare the ultrathin sections (70 to 90 nm thick) from the molded plates prepared for the mechanical testing. These sections were stained by vapor of osmium tetroxide (30 min) and ruthenium tetroxide (2 h), respectively. From the analysis of binary PS/PE and PS/SBR blends stained by this technique, PS appears as the dark gray phase, SBR as a black one, and the PE phase as light gray. In some cases, only the SBR phase was stained by osmium tetroxide. The average number of PE particles encapsulated in one SBR domain was calculated as the ratio K, between the number of PE droplets and the number of rubbery domains in which they are dispersed. (From I. Luzinov, K. Xi, C. Pagnoulle, G. Huynh-Ba, and R. Jérôme, *Polymer* 40, 2511–2520, 1999. With permission.)

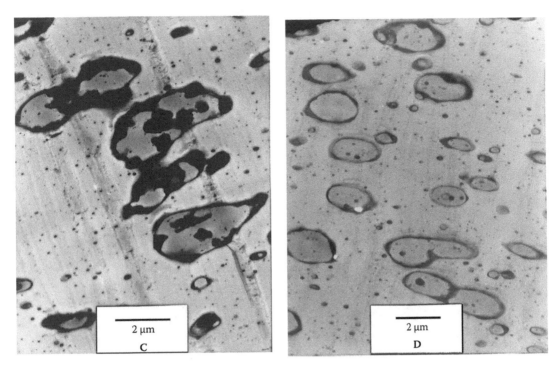

FIGURE 4.54
(*Continued*)

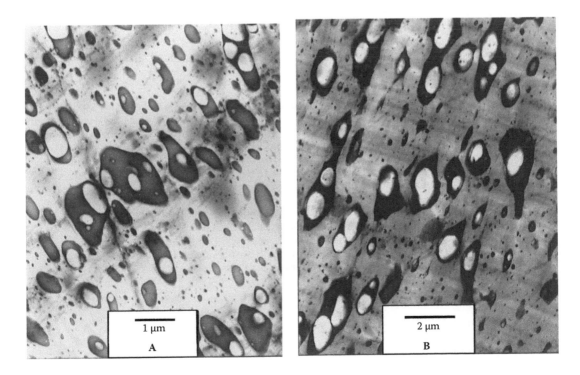

FIGURE 4.55

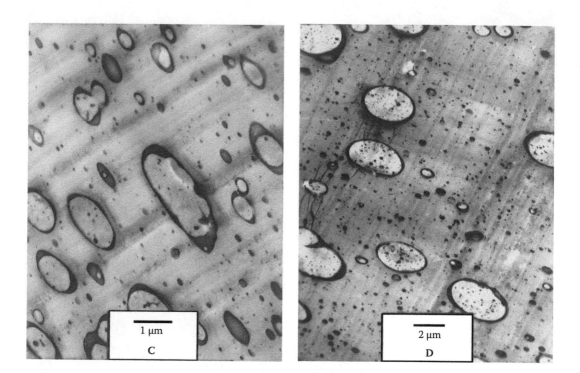

FIGURE 4.55 (*Continued*)
TEM photomicrographs of PS/SBR-2/PE blends containing all the blends investigated contain 75 wt% PS (polystyrol 158K from BASF), whereas the two minor components—that is, a rubber (SBR) and a low-density polyethylene (PE, Shell 33)—were used in various mixing ratios. SBR-1 (Shell, Cariflex S1502) and SBR-2 (Shell, Cariflex S1013) contained 23 wt% and 42.7 wt% bound styrene, respectively, of different PE contents in the mixed SBR-2/PE dispersed phase. SBR-2 has an Mw of 418 k. (A): 15%; (B): 32%; (C): 50%; and (D): 67%. The phase morphology was investigated as described in Figure 4.54. Similar to the blends consisting of SBR-1 (Figure 4.54), PE systematically forms subphases in the SBR-2 phase. At low PE contents, more than one PE subphase is usually observed in the SBR-2 domains (A, B). At PE content close to and higher than the theoretical composition of phase inversion for SBR-2/PE, the binary blends, each PE subphase is observed as coated by an SBR-2 shell (C, D). For these blends, there are subinclusions of SBR-2 in the PE phase, particularly for blends with the highest PE content. It must be pointed out that the shape of the dispersed SBR-1 and SBR-2 phases is irregular at low PE content (<50 wt%) and becomes much more regular when PE becomes the major component (Figure 4.54 and this figure). (From I. Luzinov, K. Xi, C. Pagnoulle, G. Huynh-Ba, and R. Jérôme, *Polymer* 40, 2511–2520, 1999. With permission.)

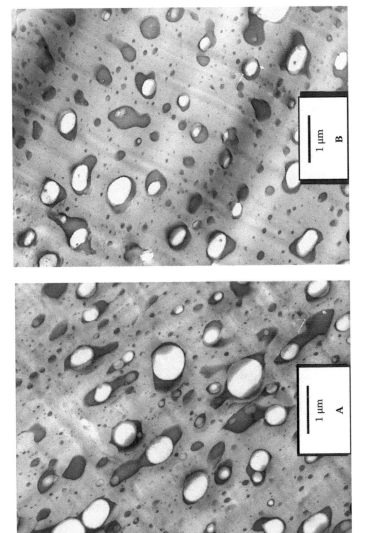

FIGURE 4.56

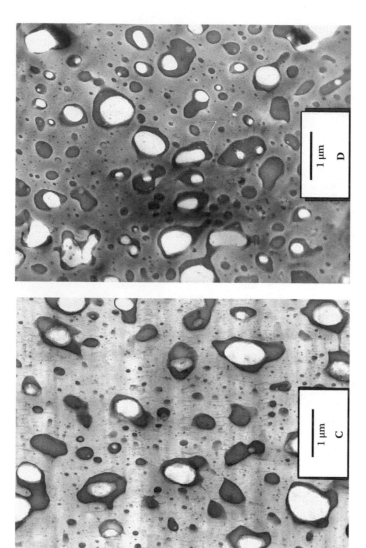

FIGURE 4.56 (*Continued*)

TEM photomicrographs for PS/SBR/PO containing 75 wt% PS, 17 wt% SBR (containing 42.7 wt% styrene), and 8 wt% PO. The blends have different polyolefins (POs). (A): poly(ethylene-co-butene) (PEBU-3); (B): low-density polyethylene (LDPE); (C): high-density polyethylene (HDPE-1); and (D): ethylene-propylene-rubber (EPR). The phase morphology was investigated as described for Figure 4.54. TEM observations confirm that the PS/SBR/PO blends have core–shell morphology, whatever the PO used. The PO phase is systematically encapsulated by SBR, as shown by TEM images for some blends consisting of different types of core-forming POs. (From I. Luzinov, K. Xi, C. Pagnoulle, G. Huynh-Ba, and R. Jérôme, *Polymer* 41, 7099–7109, 2000. With permission.)

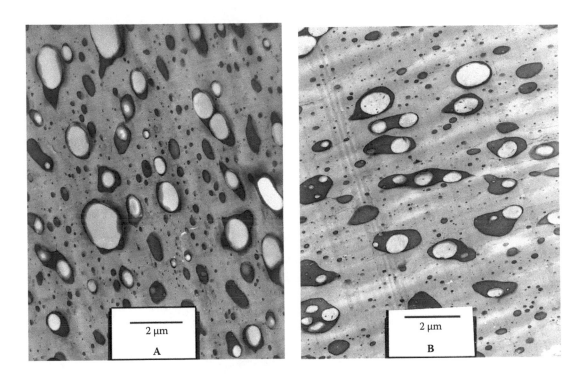

FIGURE 4.57

TEM photomicrographs of PS/SBR/PO blends containing 75 wt% PS, 17 wt% SBR (containing 42.7 wt% styrene), and 8 wt% PO. The blends have different polyolefins (POs). (A): PEBU-3; (B): LDPE; (C): HDPE-1; (D): EPR; and (E): PP. These blends were compression molded after the compounding process. Size calculation shows that the diameter of the PO particles has changed upon molding in relation to the torque of the core-forming polymer. This size actually passes through a maximum. The same general behavior is observed for the diameter of the SBR domains and the thickness of the rubbery layer around the PO subphases. As a rule, all these sizes have increased as a result of the melt pressing, more likely because of phase coarsening by coalescence. (From I. Luzinov, K. Xi, C. Pagnoulle, G. Huynh-Ba, and R. Jérôme, *Polymer* 41, 7099–7109, 2000. With permission.)

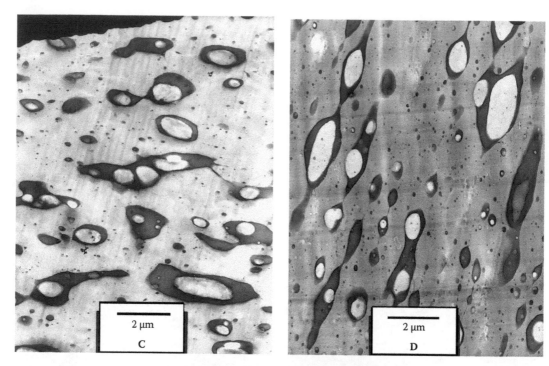

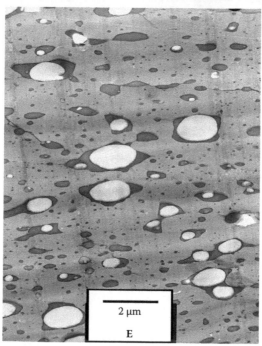

FIGURE 4.57
(*Continued*)

Single-screw extrusion; 10 wt% SEBS-g-MA

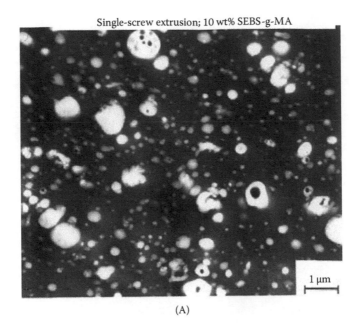

(A)

Twin-screw extrusion; 10 wt% SEBS-g-MA

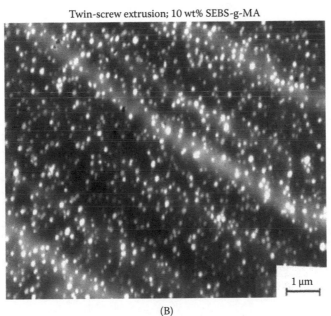

(B)

FIGURE 4.58
TEM photomicrographs of 10/90 (A, B) and 25/75 (C, D) SEBS-g-MA/a-PA blends prepared by single-screw and twin-screw extrusion. The polyamide phase is stained dark with phosphotungstic acid. The image shows typical morphological characteristics for blends with two SEBS-g-MA contents (no SEBS present) prepared in the different extruder types mentioned earlier. Unlike similar blends with nylon-66, the rubber particles in this amorphous polyamide are quite spherical and regular in shape. The extruder type does not affect the shape of the rubber particles; however, the twin-screw extruder generates much smaller particles than the single-screw extruder. Rubber particle size is seen to depend on rubber content. (From J. J. Huang, H. Keskkula, and D. R. Paul, *Polymer* 45, 4203–4215, 2004. With permission.)

Single-screw extrusion; 25 wt% SEBS-g-MA

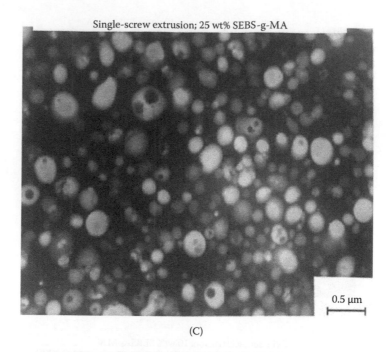

0.5 μm

(C)

Twin-screw extrusion; 25 wt% SEBS-g-MA

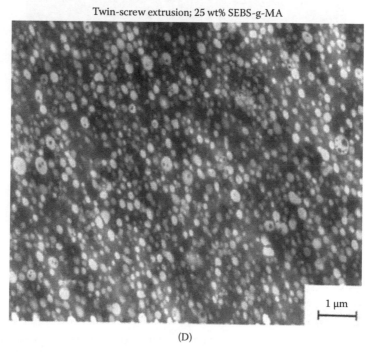

1 μm

(D)

FIGURE 4.58
(*Continued*)

Single-screw extrusion; 0/100 SEBS-g-MA/SEBS

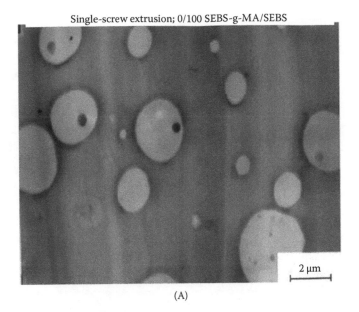

2 μm

(A)

Twin-screw extrusion; 0/100 SEBS-g-MA/SEBS

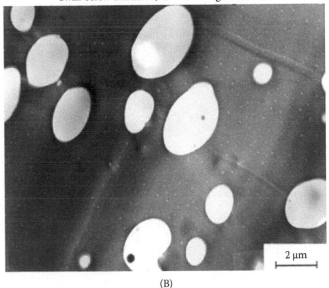

2 μm

(B)

FIGURE 4.59

TEM photomicrographs of 20 wt% rubber/80 wt% polyamide blends with the ratio of maleic anhydride grafted styrene-ethylene butylene-styrene (SEBS-g-MA)/SEBS equal to 0/100 (A, B) and 80/20 (C, D) prepared by single-screw and twin-screw extrusion. The polyamide phase is stained dark with phosphotungstic acid. These images compare the morphology of blends, where the rubber phase (20 wt%) contains 0 and 80 wt% SEBS-g-MA that were prepared by both the single- and the twin-screw extruders. Again, the rubber particles are regular in shape but the particle size for the blend containing no SEBS-g-MA is much larger than that of the blend, where the rubber phase consists of 80 wt% SEBS-g-MA. (From J. J. Huang, H. Keskkula, and D. R. Paul, *Polymer* 45, 4203–4215, 2004. With permission.)

Single-screw extrusion; 80/20 SEBS-g-MA/SEBS

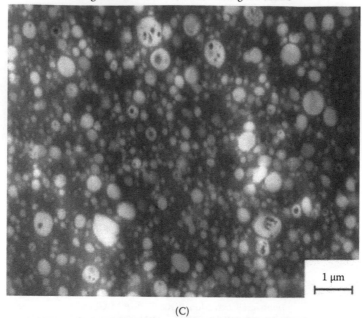

1 µm

(C)

Twin-screw extrusion; 80/20 SEBS-g-MA/SEBS

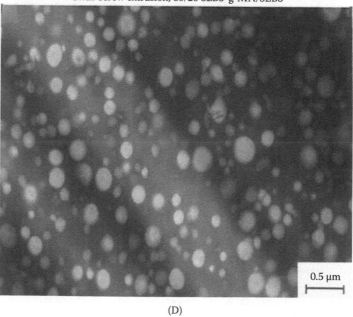

0.5 µm

(D)

FIGURE 4.59
(*Continued*)

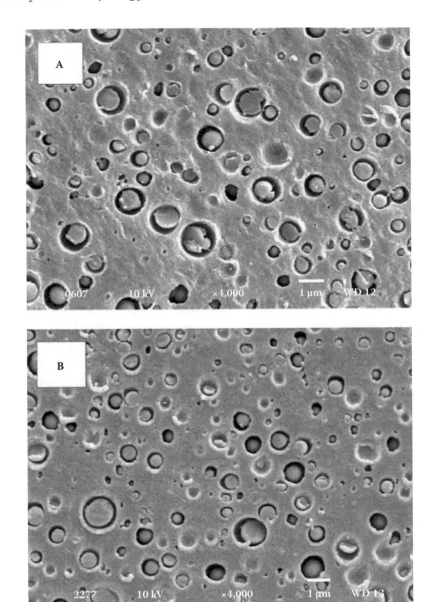

FIGURE 4.60
SEM photomicrographs of blend 1 having HDPE as a matrix in which are dispersed the two minor phases (low-Mw polystyrene/low-Mw PMMA) at 2 min (A) and 15 min (B) of mixing time. PS is extracted by cyclohexane. A stable composite droplet morphology is obtained within 2 min of mixing. The white scale bar denotes 1 μm. (From J. Reignier, B. D. Favis, and M.-Cl. Heuzey, *Polymer* 44, 49–59, 2003. With permission.)

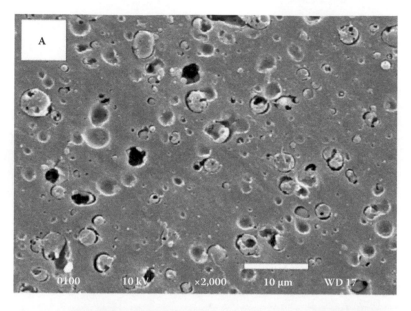

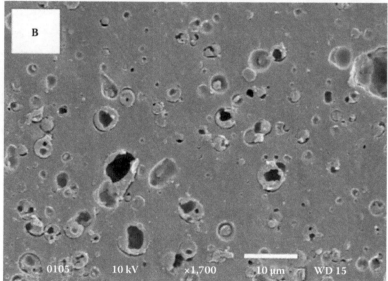

FIGURE 4.61
SEM photomicrographs of blend 3 having HDPE as a matrix in which are dispersed the two minor phases (high-Mw polystyrene/high-Mw PMMA) at 2 min (A) and 15 min (B) of mixing time. PS is extracted by cyclohexane. As mixing time progresses, a more complete encapsulation of PS by PMMA is observed. The white scale bar denotes 10 μm. The images show an encapsulation of PMMA particles by a PS shell. Significantly changing the PMMA core molecular weight has no effect on the extent, or the kinetics, of composite droplet formation. All the PMMA is located as subinclusions in the PS dispersed phase within 2 min of mixing. The analysis of the extent of composite reveals that 100% of the PMMA is present as subinclusions in the PS dispersed phase for blends 1 and 2 at all mixing times. The image also reveals that the morphology evolves from partial encapsulation to total encapsulation after 15 min of mixing. (From J. Reignier, B. D. Favis, and M.-Cl. Heuzey, *Polymer* 44, 49–59, 2003. With permission.)

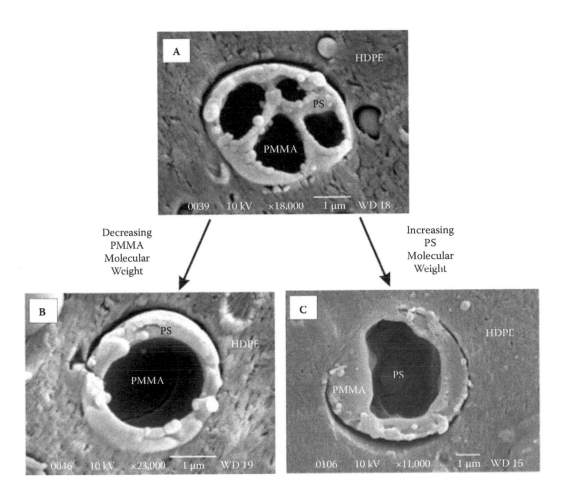

FIGURE 4.62
Dependence of the composite dispersed morphology on PS and PMMA molecular weights. SEM photomicrographs of (A) blend 1 (low-Mw PS/high-Mw PMMA), where PMMA is extracted by acetic acid and PS encapsulates PMMA; (B) blend 2 (low-Mw PS/low-Mw PMMA), where PMMA is extracted by acetic acid and PS encapsulates PMMA; and (C) blend 3 (high-Mw PS/high-Mw PMMA), where PS is extracted by cyclohexane and PMMA encapsulates PS. In all cases HDPE is the matrix. The white scale bar denotes 1 μm. The results clearly show that in blends 1 and 2, PS encapsulates PMMA and in blend 3 the PMMA encapsulates PS. (From J. Reignier, B. D. Favis, and M.-Cl. Heuzey, *Polymer* 44, 49–59, 2003. With permission.)

5

Scanning Electron Microscopy Illustrations of Two-Phase Cocontinuous Morphologies in Binary Polymer Blends

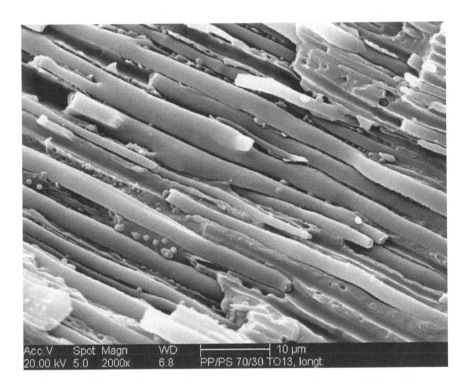

FIGURE 5.1

Scanning electron microscopy (SEM) photomicrograph (2000× magnification) of cryofracture surface of poly-propylene/polystyrene (PP/PS) blends. Another view of cylinder-like structure of polystyrene minor phase in melt-blended 70 wt% polypropylene/30 wt% polystyrene blends. This image illustrates clearly the extent of flow-induced orientation during the process of melt-extrusion. Note that some spherical particles coexist with the overall oriented structure. These can originate from breakup of fine cylinders (fibers). (From T. S. Omonov, *Crucial Aspects of Phase Morphology Generation and Stabilization in Two- and Three-Phase Polymer Blends: Physical, Reactive and Combined Routes of Compatibilization*, Ph.D. thesis, Katholieke Universiteit Leuven, Belgium, 2007, under the supervision of C. Harrats and G. Groeninckx.)

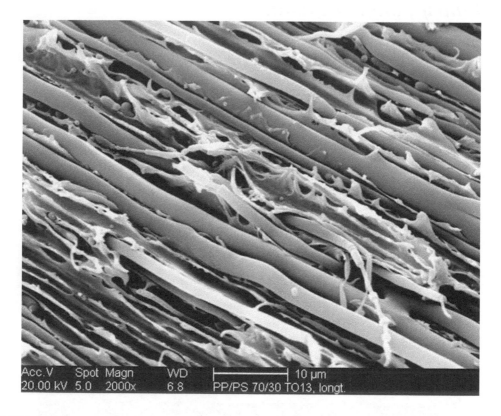

FIGURE 5.2

SEM photomicrograph (2000× magnification) showing highly oriented structure in melt-blended polypropyl-ene/polystyrene blends. Cryofracture surfaces. This is an additional example of anisotropic phase morphol-ogy in 70 wt% polypropylene/30 wt% polystyrene blends. The elongated entities of the minor phase (30 wt% polystyrene) are clearly visible in the matrix (70 wt% polypropylene). They are cylinders of infinite length as a result of extensive orientation in the die entrance during the extrusion process. Structure observed in the longitudinal direction with respect to the extrusion axis. (From T. S. Omonov, *Crucial Aspects of Phase Morphology Generation and Stabilization in Two- and Three-Phase Polymer Blends: Physical, Reactive and Combined Routes of Compatibilization*, Ph.D. thesis, Katholieke Universiteit Leuven, Belgium, 2007, under the supervision of C. Harrats and G. Groeninckx.)

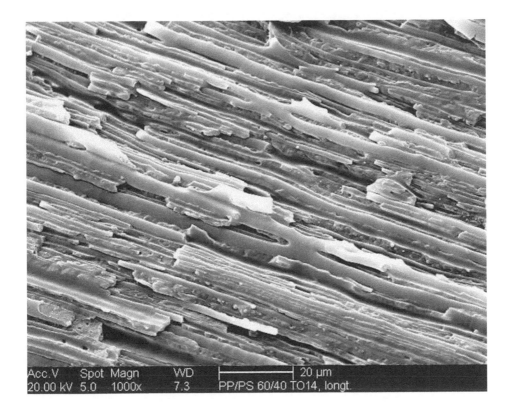

FIGURE 5.3
SEM photomicrograph (1000× magnification) of cryofracture surface of 60 wt% polypropylene/40 wt% poly-styrene. This image reveals that the cylinders are getting more in contact and the overall structure contains more flat structures, indicating the formation of a sheet-like minor phase structure. Four-way branches are clearly visible. (From T. S. Omonov, *Crucial Aspects of Phase Morphology Generation and Stabilization in Two- and Three-Phase Polymer Blends: Physical, Reactive and Combined Routes of Compatibilization*, Ph.D. thesis, Katholieke Universiteit Leuven, Belgium, 2007, under the supervision of C. Harrats and G. Groeninckx.)

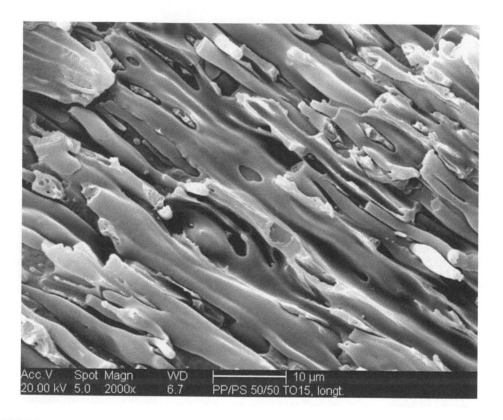

FIGURE 5.4

SEM photomicrograph of melt-blended 50 wt% polypropylene/50 wt% polystyrene blend extruded at a mixing temperature of 215°C. The PS and PP homopolymers are fed into the extruder after dry blending under nitrogen flux. The major component is introduced under nitrogen flux at a screw rotation speed of 50 rpm. As the dry-mixed components are completely fed into the extruder, the screw rotation speed is increased to 100 rpm and the blending continued for 10 min. This image illustrates how crucial is the direction and plane of observation in anisotropic phase morphologies. In this case the surface observed was broken longitudinal to the extrusion axis where the minor phase particles have been substantially elongated into elongated flat tape structures. (From T. S. Omonov, *Crucial Aspects of Phase Morphology Generation and Stabilization in Two- and Three-Phase Polymer Blends: Physical, Reactive and Combined Routes of Compatibilization*, Ph.D. thesis, Katholieke Universiteit Leuven, Belgium, 2007, under the supervision of C. Harrats and G. Groeninckx.)

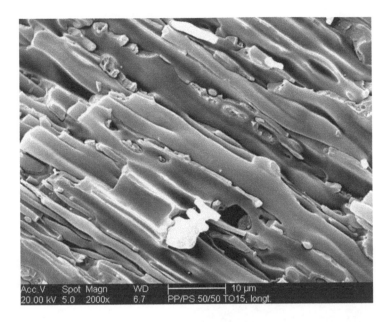

FIGURE 5.5
SEM photomicrograph of a cryofracture surface of the same sample in Figure 5.4 showing the transition between cylinder-like structure and flat sheet-like structure formation. This is, of course, due to the composition effect. The content of the minor polystyrene phase is higher compared to Figure 5.1 through Figure 5.3. (From T. S. Omonov, *Crucial Aspects of Phase Morphology Generation and Stabilization in Two- and Three-Phase Polymer Blends: Physical, Reactive and Combined Routes of Compatibilization*, Ph.D. thesis, Katholieke Universiteit Leuven, Belgium, 2007, under the supervision of C. Harrats and G. Groeninckx.)

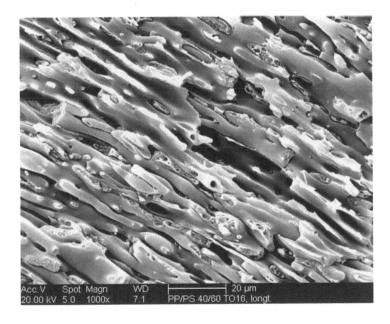

FIGURE 5.6
A much more illustrative SEM photomicrograph showing the interconnection of the matrix and the minor phase-oriented, sheet-like structure in melt-blended 40 wt% polypropylene/60 wt% polystyrene blends. (From T. S. Omonov, *Crucial Aspects of Phase Morphology Generation and Stabilization in Two- and Three-Phase Polymer Blends: Physical, Reactive and Combined Routes of Compatibilization*, Ph.D. thesis, Katholieke Universiteit Leuven, Belgium, 2007, under the supervision of C. Harrats and G. Groeninckx.)

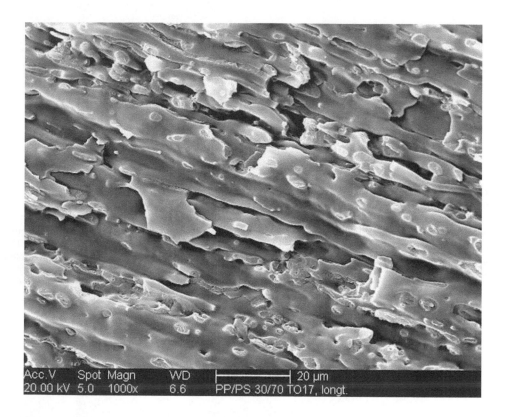

FIGURE 5.7

SEM photomicrograph (1000× magnification) of cryofracture surfaces of 30 wt% polypropylene/70 wt% polystyrene blend. The image shows another type of sheet-like cocontinuous phase morphology where the minor phase appears as sheets highly oriented in the direction of extrusion. This composition is the opposite of the samples shown in Figure 5.1 and Figure 5.2. When the polypropylene is the minor phase, it does not exhibit cylinder-like structures but flat, sheet-like entities, interconnected to form a continuum in the polystyrene matrix. (From T. S. Omonov, *Crucial Aspects of Phase Morphology Generation and Stabilization in Two- and Three-Phase Polymer Blends: Physical, Reactive and Combined Routes of Compatibilization*, Ph.D. thesis, Katholieke Universiteit Leuven, Belgium, 2007, under the supervision of C. Harrats and G. Groeninckx.)

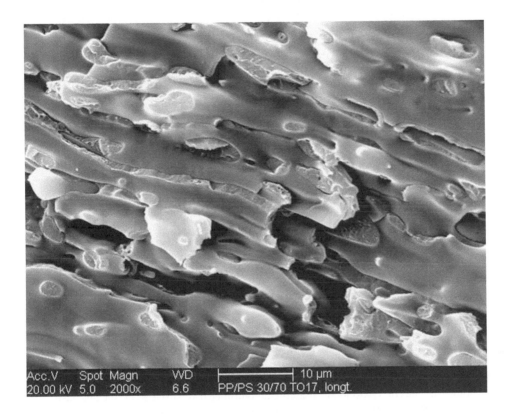

FIGURE 5.8
SEM photomicrograph (2000× magnification) of the same sample as in Figure 5.7. The image shows more details on a smaller area of observation. Note that the visible particles constitute the grafting zones by which cocontinuity is ensured over the whole blend volume. (From T. S. Omonov, *Crucial Aspects of Phase Morphology Generation and Stabilization in Two- and Three-Phase Polymer Blends: Physical, Reactive and Combined Routes of Compatibilization*, Ph.D. thesis, Katholieke Universiteit Leuven, Belgium, 2007, under the supervision of C. Harrats and G. Groeninckx.)

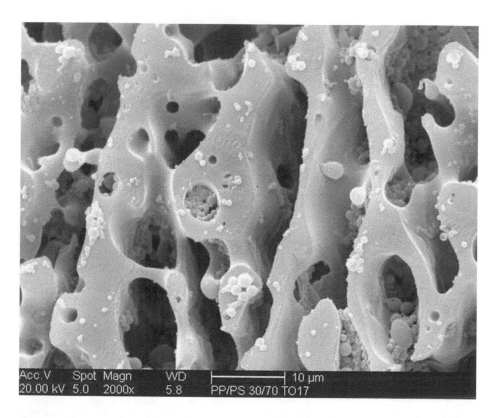

FIGURE 5.9
SEM photomicrograph of cryosmoothed and chloroform-etched surfaces of melt-blended 30 wt% of polypropylene/70 wt% polystyrene blend. This image shows complex phase morphology, although it appears simply as a common cocontinuous. A close observation inside the voids, which are the empty spaces left after the selective extraction of the polystyrene phase, reveals that the morphology of polypropylene is composed of two structures: a major continuous and tortuous phase in which are attached grape-like particles of the same phase (polypropylene). This phenomenon is quite expected, as the content of polypropylene is only 30 wt%, which makes it more vulnerable to extension and breakup. These particles remain attached as shown to the mother phase of polypropylene, which constitutes the major part of the network after the etching of the polystyrene phase. (From T. S. Omonov, *Crucial Aspects of Phase Morphology Generation and Stabilization in Two- and Three-Phase Polymer Blends: Physical, Reactive and Combined Routes of Compatibilization*, Ph.D. thesis, Katholieke Universiteit Leuven, Belgium, 2007, under the supervision of C. Harrats and G. Groeninckx.)

FIGURE 5.10
SEM photomicrograph (4000× magnification) of the same sample as in Figure 5.9, but the scale of observation is much smaller. The image shows clearly how the particles are organized in grapes that are attached to the polypropylene mother phase. If they were loose, they would be extracted as solid suspensions with the solvent of polystyrene during the etching process. (From T. S. Omonov, *Crucial Aspects of Phase Morphology Generation and Stabilization in Two- and Three-Phase Polymer Blends: Physical, Reactive and Combined Routes of Compatibilization*, Ph.D. thesis, Katholieke Universiteit Leuven, Belgium, 2007, under the supervision of C. Harrats and G. Groeninckx.)

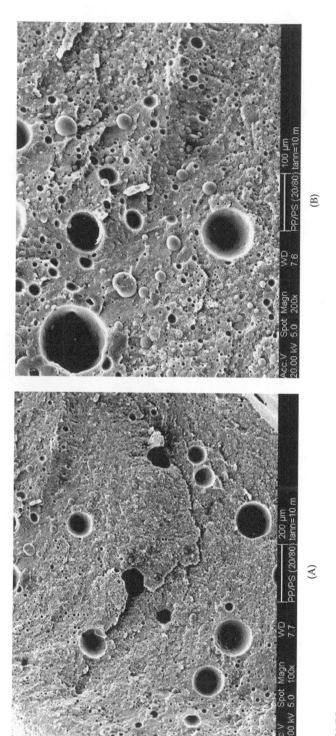

FIGURE 5.11

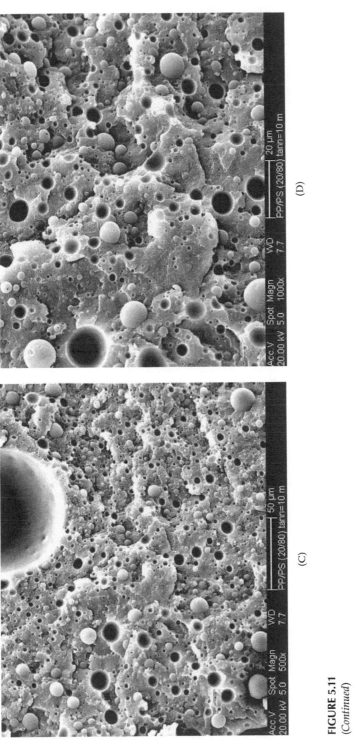

FIGURE 5.11
(*Continued*)

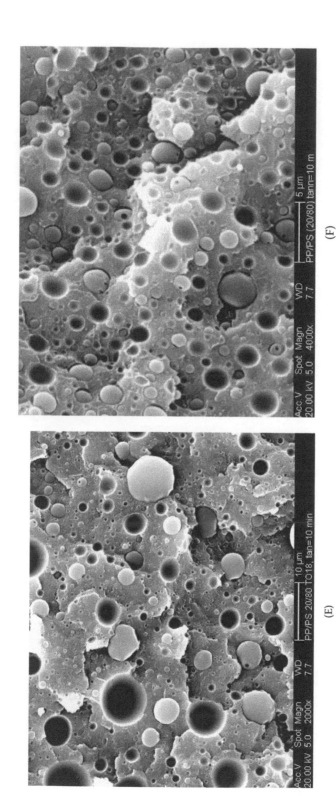

(E)   (F)

FIGURE 5.11 (*Continued*)

SEM photomicrographs of cryofracture surfaces of melt-blended 20 wt% polypropylene/80 wt% polystyrene. (A): 100× magnification; (B): 200× magnification; (C): 500× magnification; (D): 1000× magnification; (E): 2000× magnification; and (F): 4000× magnification. These images demonstrate the importance of the scale at which phase morphologies are observed and characterized. In (A) no details are revealed but heterogeneity is depicted due to the presence of a few very large particles within the range of 50 to 100 μm. Larger magnification reveals that the true particles of the minor polypropylene phase are not the big observed particles but rather homogeneously dispersed smaller particles, the average size of which is as small as 2 μm as the most magnified scale shows in (F). This situation suggests that it is more prudent to present few images of various magnifications for each blend investigated, particularly when the phase morphologies are complex. Note that these blends are observed in the transverse to the extrusion direction. (From T. S. Omonov, *Crucial Aspects of Phase Morphology Generation and Stabilization in Two- and Three-Phase Polymer Blends: Physical, Reactive and Combined Routes of Compatibilization*, Ph.D. thesis, Katholieke Universiteit Leuven, Belgium, 2007, under the supervision of C. Harrats and G. Groeninckx.)

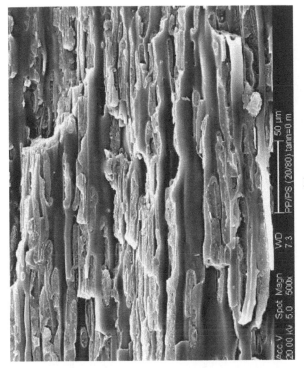

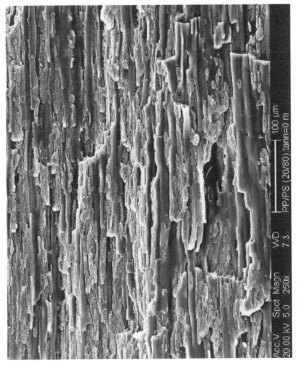

FIGURE 5.12

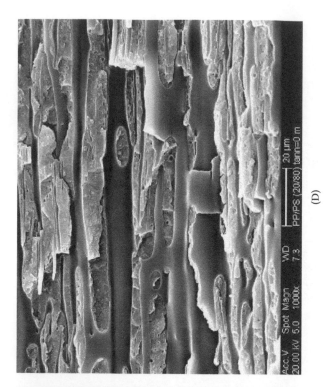

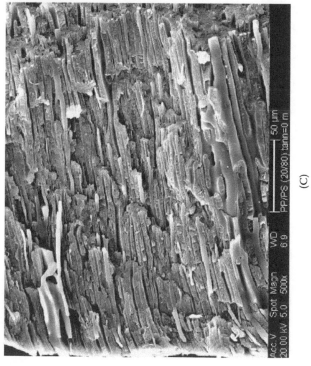

FIGURE 5.12
(Continued)

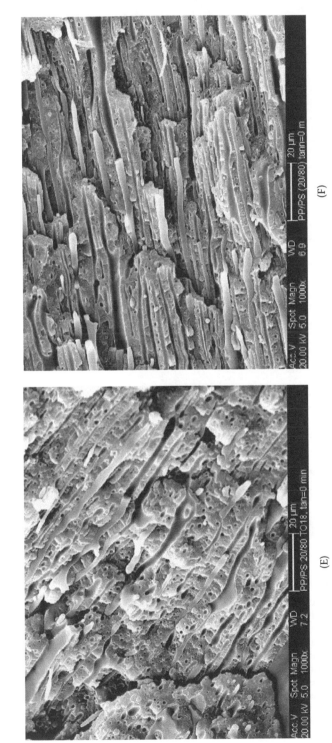

(E)

(F)

FIGURE 5.12 (*Continued*)

SEM photomicrographs of cryofracture surface of 20 wt% polypropylene/80 wt% polystyrene observed in the longitudinal direction with respect to the axis of extrusion. The sample was quenched in water after the exit from the die. It is not easy to distinguish between the two phases at this scale of observation below 1000×; however, the oriented configuration of the cylinder-like structure can be seen. (A): 250× magnification. (B): 500× magnification; at this scale a trace of the minor phase is visible as indicated by the smooth surfaces resulting from the action of cryofracturing the sample and decohesion of the polypropylene from the polystyrene matrix. The characteristics shown here do not reveal the overall features of the phase morphology of the blend. (C): 500× magnification, taken from a different area. This view provides more details. Some elongated structures of the polypropylene phase are visible. (D): 1000× magnification. At this magnification many interesting features are revealed. The polypropylene phase consists of highly elongated cylinder-like structures. Note that, to ensure phase cocontinuity, the long cylinders are bridged in the transverse direction. (E): 1000× magnification. This image is taken from another area where the elongated cylinders are clearly visible. Note also the existence of very fine perpendicular structures (about a micron in diameter). These cylinders are not visible directly, but the voids they left after the breaking of the sample are well distributed over the whole surface of observation. (F): 1000× magnification. This view is the most revealing in terms of phase morphology features. The elongated structure is clearly seen, with the transverse entities leaving visible voids of very fine dimensions. Note that the samples observed in the above figures exhibit cocontinuous phase morphology, which is not stable upon the thermal annealing process at a fixed temperature of 200°C. (From T. S. Omonov, *Crucial Aspects of Phase Morphology Generation and Stabilization in Two- and Three-Phase Polymer Blends: Physical, Reactive and Combined Routes of Compatibilization*, Ph.D. thesis, Katholieke Universiteit Leuven, Belgium, 2007, under the supervision of C. Harrats and G. Groeninckx.)

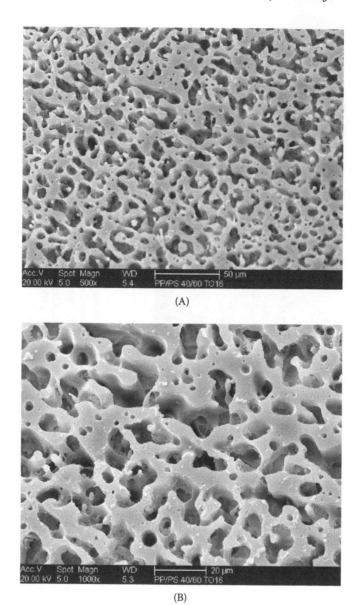

(A)

(B)

FIGURE 5.13

SEM photomicrographs of cryosmoothed surface of melt-blended 40 wt% polypropylene/60 wt% polystyrene blends. The polystyrene phase has been selectively etched using chloroform. (A): 500× magnification. This scale of observation allows a global view of the phase morphology of the polypropylene phase. A closer examination of the image reveals some lack of homogeneity mainly on the top-center of the image where the forms are gross. (B): 1000× magnification. At this scale the image shows more details—mainly spherical voids, which are synonymous of the existence of cylinder-like structures of polystyrene embedded within the coral-like continuous phase of the polypropylene. Also the fine structures of the polypropylene phase in the form of cylinders or spheres are attached to the mother PP phase in the form of grapes. (C): 2000× magnification. This is a clear illustration of (B). The droplets of PP in the form of grapes appear grafted to the PP host coral-like continuous structure. (D): 4000× magnification. This is the largest magnification above which no interesting information on phase morphology can be obtained. (From T. S. Omonov, *Crucial Aspects of Phase Morphology Generation and Stabilization in Two- and Three-Phase Polymer Blends: Physical, Reactive and Combined Routes of Compatibilization*, Ph.D. thesis, Katholieke Universiteit Leuven, Belgium, 2007, under the supervision of C. Harrats and G. Groeninckx.)

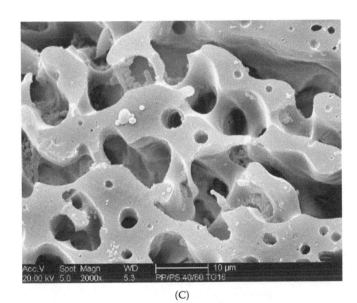

(C)

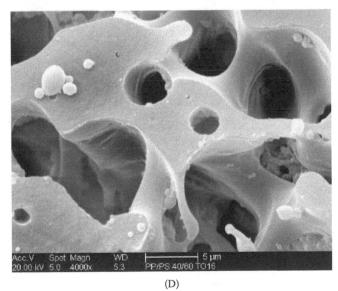

(D)

FIGURE 5.13
(*Continued*)

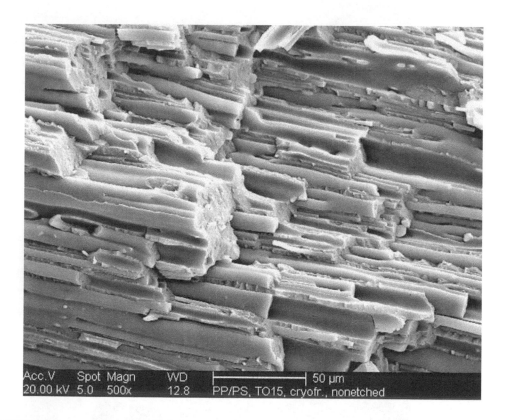

FIGURE 5.14

SEM photomicrograph of cryofracture nonetched surface of melt-blended 50 wt% polystyrene/50 wt% polypropylene blends; magnification of 500×. The sample was cut in the longitudinal direction with respect to the flow of extrusion. This image, which reveals a nonuniformly broken surface, shows two shells of structures. The elongated structures (cylinder-like shape) of the polystyrene phase are evident, embedded in the polypropylene phase. Because the composition of the blend is symmetrical (50/50), the two phases appear almost to have the same elongated structures, one inside the other, but a close examination reveals that the PS phase is the dispersion and the polypropylene the matrix (as confirmed by selective extraction). (From T. S. Omonov, *Crucial Aspects of Phase Morphology Generation and Stabilization in Two- and Three-Phase Polymer Blends: Physical, Reactive and Combined Routes of Compatibilization*, Ph.D. thesis, Katholieke Universiteit Leuven, Belgium, 2007, under the supervision of C. Harrats and G. Groeninckx.)

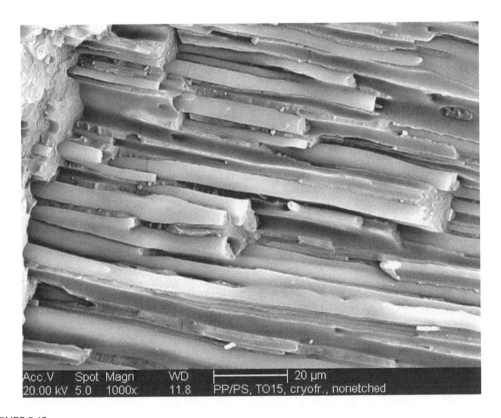

FIGURE 5.15

SEM photomicrograph of cryofracture surface, nonetched, of melt-blended 50 wt% polystyrene/50 wt% polypropylene blends; magnification of 1000×. The sample was cut in the longitudinal direction with respect to the flow of extrusion. This image of a surface uniformly broken and of higher magnification shows clearly how the polystyrene cylinders are well aligned and firmly embedded in the polypropylene matrix. A close observation of the photomicrograph reveals the existence on the polystyrene cylinders of particles, which are graft points to ensure three-dimensional cocontinuous structures. Note also that the polystyrene cylinders are interconnected in four-way branches. This full cocontinuity has also been demonstrated by selective solvent extraction and weighing of the polystyrene phase using chloroform. (From T. S. Omonov, *Crucial Aspects of Phase Morphology Generation and Stabilization in Two- and Three-Phase Polymer Blends: Physical, Reactive and Combined Routes of Compatibilization*, Ph.D. thesis, Katholieke Universiteit Leuven, Belgium, 2007, under the supervision of C. Harrats and G. Groeninckx.)

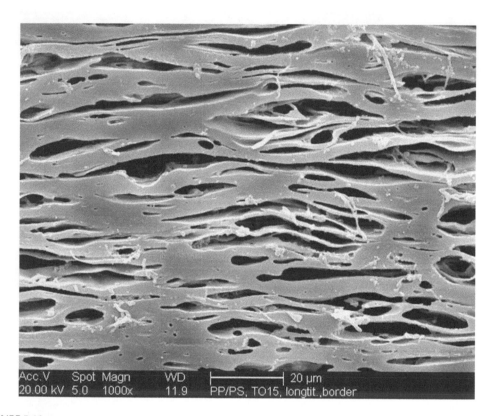

FIGURE 5.16
SEM photomicrograph of cryosmoothed and chloroform-etched surface of melt-blended 50 wt% polystyrene/
50 wt% polypropylene blends; magnification of 1000×. The sample was cut in the longitudinal direction with
respect to the flow of extrusion. The surface is cryosmoothed using a diamond knife and the polystyrene phase
etched using chloroform. Here the artifact is the collapsing of the remaining (after etching) PP phase. It mis-
leads about the exact size of the elongated cylinders. In this particular case and for this type of phase morphology the examination of the cryofracture surfaces is preferred to the examination of the cryosmoothed ones.
Cryofracturing is a less destructive technique than cryosmoothing. (From T. S. Omonov, *Crucial Aspects of Phase
Morphology Generation and Stabilization in Two- and Three-Phase Polymer Blends: Physical, Reactive and Combined
Routes of Compatibilization*, Ph.D. thesis, Katholieke Universiteit Leuven, Belgium, 2007, under the supervision of
C. Harrats and G. Groeninckx.)

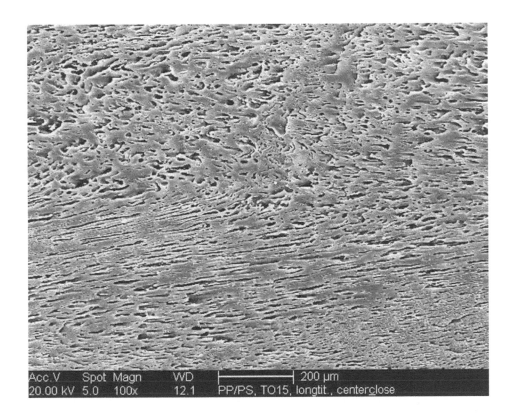

FIGURE 5.17
SEM photomicrograph of cryosmoothed and chloroform-etched surfaces of melt-blended 50 wt% polystyrene/ 50 wt% polypropylene blends observed at very low magnification of 100×. The sample was cut in the longitudinal direction with respect to the flow of extrusion. This image reveals that selective phase etching can be useful in showing how homogeneous the minor phase structure is in contrast to the cryofracture, which, depending on the blend composition, can lead to lack of characterization. Indeed, heterogeneous phase morphology of the minor phase is revealed in this blend sample. It contains both elongated and flat nonuniform structures. This is due to the processing conditions. (From T. S. Omonov, *Crucial Aspects of Phase Morphology Generation and Stabilization in Two- and Three-Phase Polymer Blends: Physical, Reactive and Combined Routes of Compatibilization*, Ph.D. thesis, Katholieke Universiteit Leuven, Belgium, 2007, under the supervision of C. Harrats and G. Groeninckx.)

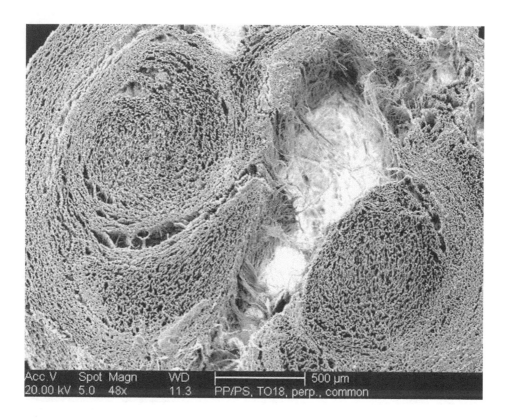

FIGURE 5.18
SEM photomicrograph (magnification of 48×) of cryofracture, nonsmoothed, but chloroform-etched surfaces of melt-blended 20 wt% polypropylene/80 wt% polystyrene blends observed at very small scale. This image shows how an extrudate looks when observed at very large surface (low magnification). The diameter of the extrudate as obtained using a midiextruder is about 4 to 5 mm. Here the heterogeneity is clear; similar morphology can be seen at the two tips of the two screws of the mixing extruder. All around the center point a gradient of change in phase morphology is visible. This confirms how "dangerous" a careless sampling can be when observing the morphology of melt-compounded blends using extruders. A very large-scale observation is necessary as a prime guideline for sampling. (The polystyrene phase was selectively etched using chloroform.) (From T. S. Omonov, *Crucial Aspects of Phase Morphology Generation and Stabilization in Two- and Three-Phase Polymer Blends: Physical, Reactive and Combined Routes of Compatibilization*, Ph.D. thesis, Katholieke Universiteit Leuven, Belgium, 2007, under the supervision of C. Harrats and G. Groeninckx.)

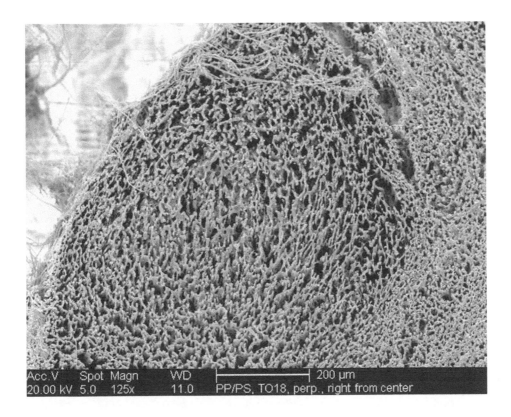

FIGURE 5.19
SEM photomicrograph of cryofracture, nonsmoothed, but chloroform-etched surfaces of melt-blended 20 wt% polypropylene/80 wt% polystyrene blends illustrating the phase morphology at the tip of the screw of the extruder. Right lobe of the two lobes shown in Figure 5.18 observed at higher magnification of 125x. (From T. S. Omonov, *Crucial Aspects of Phase Morphology Generation and Stabilization in Two- and Three-Phase Polymer Blends: Physical, Reactive and Combined Routes of Compatibilization*, Ph.D. thesis, Katholieke Universiteit Leuven, Belgium, 2007, under the supervision of C. Harrats and G. Groeninckx.)

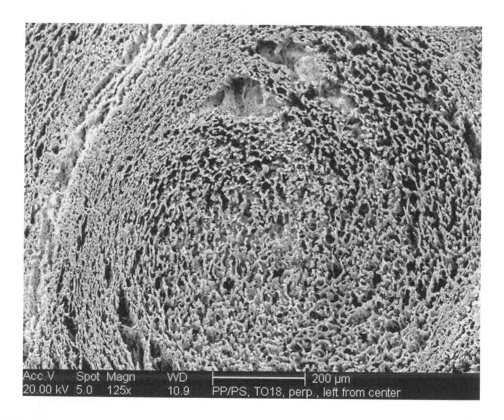

FIGURE 5.20
SEM photomicrograph of cryofracture, nonsmoothed, but chloroform-etched surfaces of melt-blended 20 wt% polypropylene/80 wt% polystyrene blends illustrating the phase morphology at the tip of the screw of the extruder. Left lobe of the two lobes shown in Figure 5.18 observed at higher magnification of 125×. (From T. S. Omonov, *Crucial Aspects of Phase Morphology Generation and Stabilization in Two- and Three-Phase Polymer Blends: Physical, Reactive and Combined Routes of Compatibilization*, Ph.D. thesis, Katholieke Universiteit Leuven, Belgium, 2007, under the supervision of C. Harrats and G. Groeninckx.)

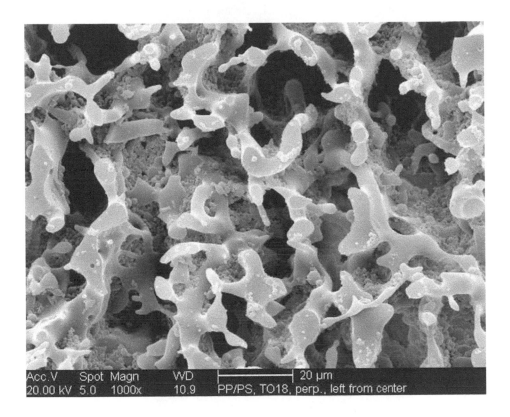

In-image text: Acc.V Spot Magn WD 20 µm
20.00 kV 5.0 1000x 10.9 PP/PS, TO18, perp., left from center

FIGURE 5.21

SEM photomicrograph of cryofracture, nonsmoothed, but chloroform-etched surfaces of melt-blended 20 wt% polypropylene/80 wt% polystyrene blends illustrating the phase morphology at the tip of the screw of the extruder. The image is taken from the right lobe of the two lobes shown in Figure 5.18 observed at much larger magnification of 1000×. The image illustrates the homogeneous cocontinuous coral-like phase morphology of the blend phases. (From T. S. Omonov, *Crucial Aspects of Phase Morphology Generation and Stabilization in Two- and Three-Phase Polymer Blends: Physical, Reactive and Combined Routes of Compatibilization*, Ph.D. thesis, Katholieke Universiteit Leuven, Belgium, 2007, under the supervision of C. Harrats and G. Groeninckx. For detailed scientific discussion refer to T. S. Omonov, C. Harrats, G. Groeninckx, and P. Moldenaers, *Polymer* 48, 5289, 2007.)

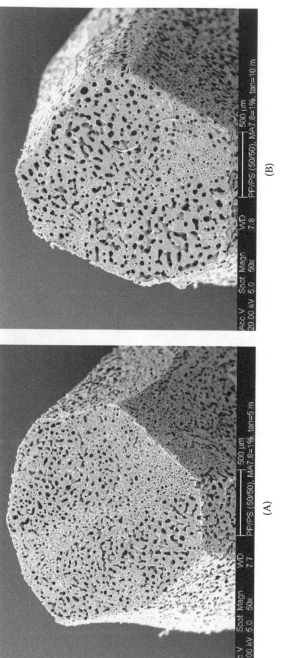

FIGURE 5.22

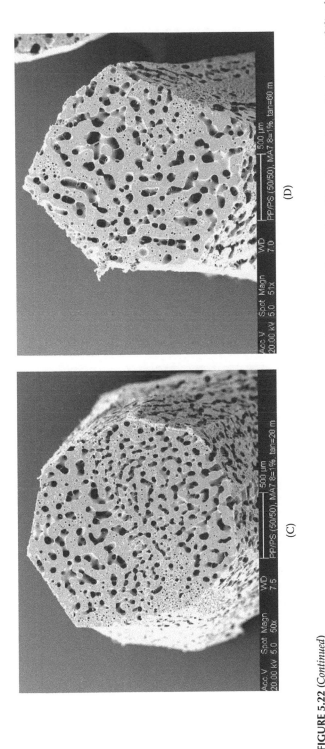

(C)

(D)

FIGURE 5.22 *(Continued)*

SEM photomicrographs of cryosmoothed and chloroform-etched surface of melt-blended 50 wt% polypropylene/50 wt% polystyrene blend compatibilized using 1 wt% of maleic anhydride grafted polypropylene (containing 7.8 wt% of maleic anhydride)/amine end-capped polystyrene copolymers thermally annealed at 200°C for various time periods. (A): 5 min. The low magnification of 50× shows the overall phase morphology of the blend. The blend exhibits mixed phase morphology composed of coral-like cocontinuous and dispersed droplets too. (B): 10 min; the image reveals a significant phase coarsening due to thermal annealing. (C): 20 min; a further increase in phase size of the coral-like structure is observed after 20 min thermal annealing. (D): 60 min; a much larger phase coarsening is obtained upon thermal annealing during a time period of 60 min. (From T. S. Omonov, *Crucial Aspects of Phase Morphology Generation and Stabilization in Two- and Three-Phase Polymer Blends: Physical, Reactive and Combined Routes of Compatibilization,* Ph.D. thesis, Katholieke Universiteit Leuven, Belgium, 2007, under the supervision of C. Harrats and G. Groeninckx. For a detailed scientific discussion refer to T. S. Omonov, C. Harrats, G. Groeninckx, and P. Moldenaers, *Polymer* 48, 5289, 2007.)

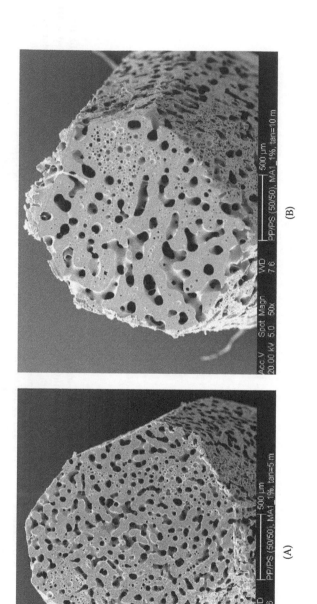

FIGURE 5.23

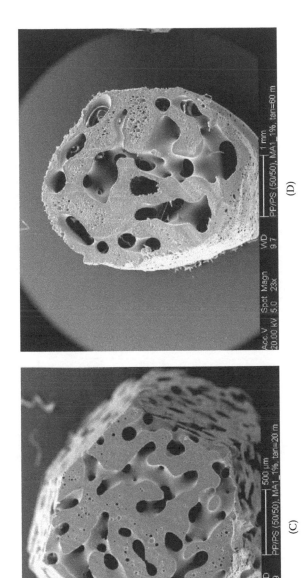

(D)

(C)

FIGURE 5.23 *(Continued)*
SEM photomicrographs (magnification 50x) of the cryosmoothed and chloroform-etched surface of melt-blended 50 wt% polypropylene/50 wt% polystyrene blend compatibilized using 1 wt% of maleic anhydride grafted polypropylene (containing 1 wt% of maleic anhydride)/amine end-capped polystyrene copolymers. The blends were thermally annealed at 200°C for various time periods to study the efficiency of the compatibilizer in stabilizing the phase morphology upon thermal annealing. (A): 5 min; the phase size is much larger than in Figure 5.22A because of the lack of compatibilizing and stabilizing efficiency of PP-MA1 in comparison to PP-MA8. This issue has been thoroughly discussed elsewhere (see cited reference). (B): 10 min. A substantial phase coarsening is observed as a result of longer annealing time. (C): 20 min. This image illustrates clearly the extent of phase coarsening, which indicates the poor stabilizing ability of PP-MA1–based compatibilizing agent. (D): 60 min. This photomicrograph observed at lower magnification (23x) illustrates the phase coarsening at a millimeter scale resulting from annealing the blend during 1 h at 200°C. Note that the dispersed polystyrene particles (holes) do not exhibit coarsening, which is indicative of their isolated state in the polypropylene matrix. (From T. S. Omonov, *Crucial Aspects of Phase Morphology Generation and Stabilization in Two- and Three-Phase Polymer Blends: Physical, Reactive and Combined Routes of Compatibilization*, Ph.D. thesis, Katholieke Universiteit Leuven, Belgium, 2007, under the supervision of C. Harrats and G. Groeninckx. For a detailed scientific discussion refer to T. S. Omonov, C. Harrats, G. Groeninckx, and P. Moldenaers, *Polymer* 48, 5289, 2007.)

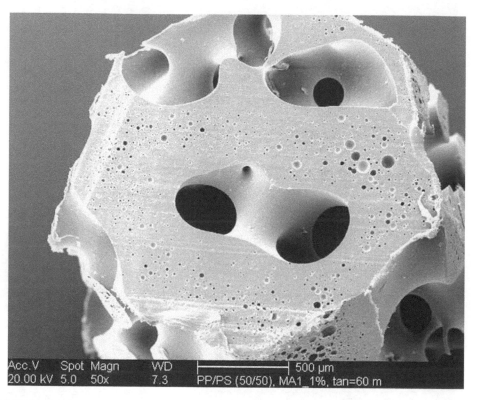

FIGURE 5.24

SEM photomicrograph (magnification 50×) of the cryosmoothed and chloroform-etched surface of melt-blended 50 wt% polypropylene/50 wt% polystyrene blend thermally annealed for 10 min at 200°C. The blend was compatibilized using 1 wt% of maleic anhydride grafted polypropylene (containing 1 wt% of maleic anhydride)/ amine end-capped polystyrene copolymers. The magnification of 50× shows the extensive phase coarsening of the blend exhibit and reveals clearly the existence of a minor part of droplet-in-matrix phase morphology in addition to the coral-like cocontinuous one. (T. S. Omonov, *Crucial Aspects of Phase Morphology Generation and Stabilization in Two- and Three-Phase Polymer Blends: Physical, Reactive and Combined Routes of Compatibilization*, Ph.D. thesis, Katholieke Universiteit Leuven, Belgium, 2007, under the supervision of C. Harrats and G. Groeninckx. For a detailed scientific discussion refer to T. S. Omonov, C. Harrats, G. Groeninckx, and P. Moldenaers, *Polymer* 48, 5289, 2007.)

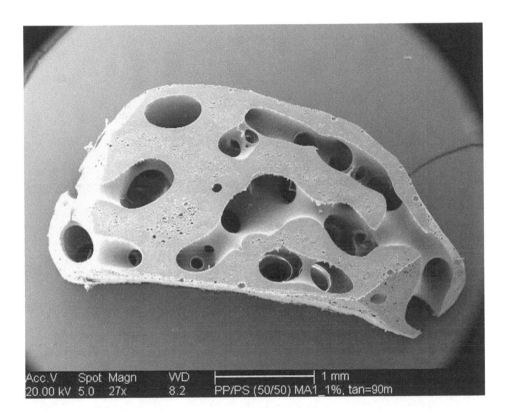

FIGURE 5.25

SEM photomicrograph (magnification 27×) of cryosmoothed and chloroform-etched surface of melt-blended 50 wt% polypropylene/50 wt% polystyrene blend thermally annealed for 90 min at 200°C. The blend was compatibilized using 1 wt% of maleic anhydride grafted polypropylene (containing 1 wt% of maleic anhydride)/amine end-capped polystyrene copolymers. After 1 h 30 min of annealing, the blend exhibits a millimeter-size phase coarsening. The tortuous pattern of the cocontinuous phase morphology has been extensively collapsed due to the flow-driven phase coarsening. Note that at this stage of evolution the blend could maintain full phase continuity as demonstrated by the selective etching of the polystyrene phase using chloroform. (From T. S. Omonov, *Crucial Aspects of Phase Morphology Generation and Stabilization in Two- and Three-Phase Polymer Blends: Physical, Reactive and Combined Routes of Compatibilization*, Ph.D. thesis, Katholieke Universiteit Leuven, Belgium, 2007, under the supervision of C. Harrats and G. Groeninckx. For a detailed scientific discussion refer to T. S. Omonov, C. Harrats, G. Groeninckx, and P. Moldenaers, *Polymer* 48, 5289, 2007.)

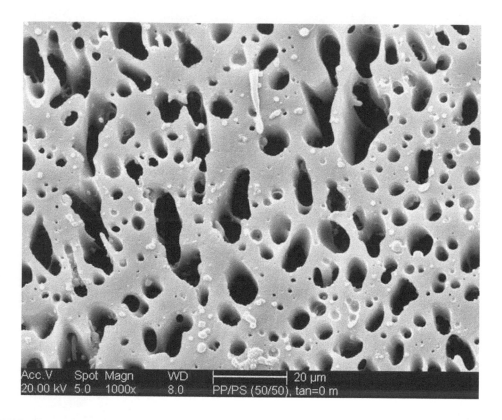

FIGURE 5.26
SEM photomicrograph (magnification 1000×) of cryosmoothed and chloroform-etched surfaces of uncompatibilized and nonannealed melt-blended 50 wt% polypropylene/50 wt% polystyrene blend. (From T. S. Omonov, *Crucial Aspects of Phase Morphology Generation and Stabilization in Two- and Three-Phase Polymer Blends: Physical, Reactive and Combined Routes of Compatibilization,* Ph.D. thesis, Katholieke Universiteit Leuven, Belgium, 2007, under the supervision of C. Harrats and G. Groeninckx.)

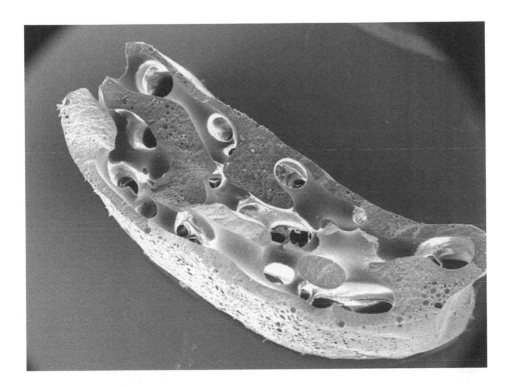

FIGURE 5.27

SEM photomicrograph (magnification 27×) of cryofractured and chloroform-etched surface of melt-blended 50 wt% polypropylene/50 wt% polystyrene blend thermally annealed for 90 min at 200°C. The blend was compatibilized using 1 wt% of maleic anhydride grafted polypropylene (containing 1 wt% of maleic anhydride)/amine end-capped polystyrene copolymers. This sample has not been cryosmoothed but only cryofractured and etched to remove polystyrene using chloroform. The cryofracture shows better the existence of isolated droplets of polystyrene in the polypropylene phase. (From T. S. Omonov, *Crucial Aspects of Phase Morphology Generation and Stabilization in Two- and Three-Phase Polymer Blends: Physical, Reactive and Combined Routes of Compatibilization*, Ph.D. thesis, Katholieke Universiteit Leuven, Belgium, 2007, under the supervision of C. Harrats and G. Groeninckx. For a detailed scientific discussion refer to T. S. Omonov, C. Harrats, G. Groeninckx, and P. Moldenaers, *Polymer* 48, 5289, 2007.)

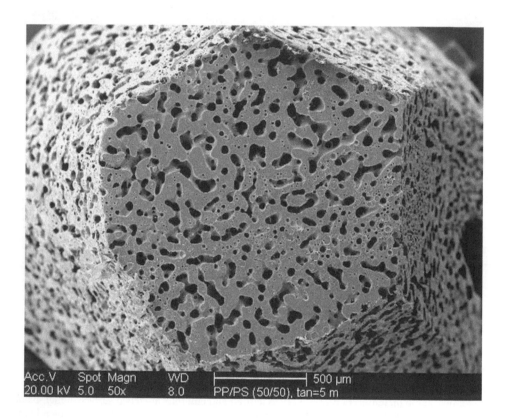

FIGURE 5.28

SEM photomicrograph (magnification 50×) of cryofractured and chloroform-etched surface of melt-blended uncompatibilized 50 wt% polypropylene/50 wt% polystyrene blend thermally annealed for 5 min at 200°C. The image reveals that in the absence of compatibilizer the phase coarsening starts after only 5 min of annealing. (From T. S. Omonov, *Crucial Aspects of Phase Morphology Generation and Stabilization in Two- and Three-Phase Polymer Blends: Physical, Reactive and Combined Routes of Compatibilization*, Ph.D. thesis, Katholieke Universiteit Leuven, Belgium, 2007, under the supervision of C. Harrats and G. Groeninckx.)

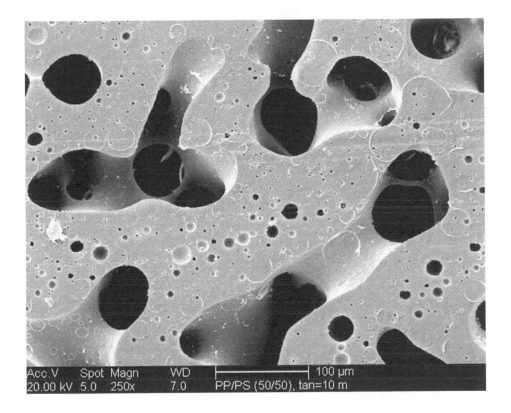

FIGURE 5.29

SEM photomicrograph (magnification 250×) of cryofractured and chloroform-etched surface of melt-blended uncompatibilized 50 wt% polypropylene/50 wt% polystyrene blend thermally annealed for 10 min at 200°C. (From T. S. Omonov, *Crucial Aspects of Phase Morphology Generation and Stabilization in Two- and Three-Phase Polymer Blends: Physical, Reactive and Combined Routes of Compatibilization*, Ph.D. thesis, Katholieke Universiteit Leuven, Belgium, 2007, under the supervision of C. Harrats and G. Groeninckx.)

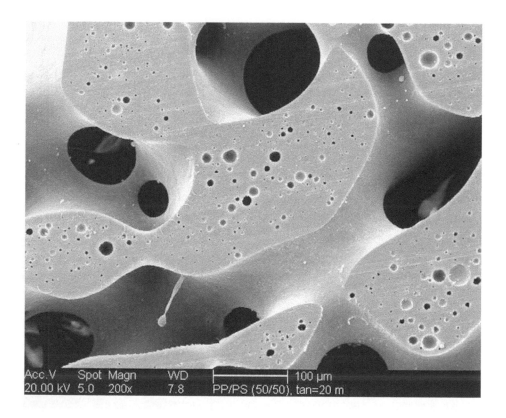

FIGURE 5.30
SEM photomicrograph (magnification 200×) of cryosmoothed and chloroform-etched surface of melt-blended uncompatibilized 50 wt% polypropylene/50 wt% polystyrene blend thermally annealed for 20 min at 200°C. Substantial phase coarsening resulting from the instability of the cocontinuous phase morphology in the absence of efficient compatibilizer is evidenced by this image. (From T. S. Omonov, *Crucial Aspects of Phase Morphology Generation and Stabilization in Two- and Three-Phase Polymer Blends: Physical, Reactive and Combined Routes of Compatibilization*, Ph.D. thesis, Katholieke Universiteit Leuven, Belgium, 2007, under the supervision of C. Harrats and G. Groeninckx.)

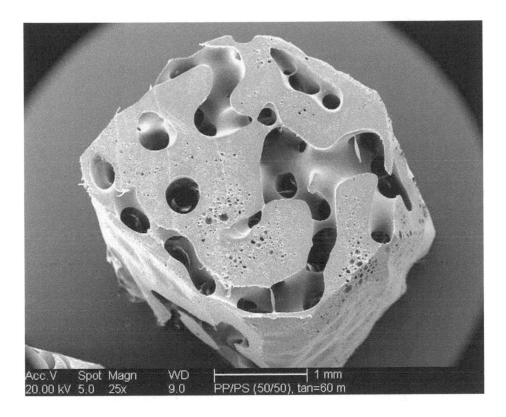

FIGURE 5.31
SEM photomicrograph (magnification 25×) of cryosmoothed and chloroform-etched surface of melt-blended uncompatibilized 50 wt% polypropylene/50 wt% polystyrene blend thermally annealed for 60 min at 200°C. Very gross coarsening has been reached after 1 h of thermal annealing in the absence of compatibilization. Most of the polystyrenes of the continuous parts are communicating on a millimeter scale. (From T. S. Omonov, *Crucial Aspects of Phase Morphology Generation and Stabilization in Two- and Three-Phase Polymer Blends: Physical, Reactive and Combined Routes of Compatibilization*, Ph.D. thesis, Katholieke Universiteit Leuven, Belgium, 2007, under the supervision of C. Harrats and G. Groeninckx.)

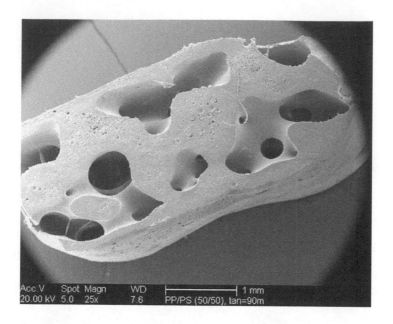

FIGURE 5.32

SEM photomicrograph (magnification 25×) of cryosmoothed and chloroform-etched surface of melt-blended uncompatibilized 50 wt% polypropylene/50 wt% polystyrene blend thermally annealed for 90 min at 200°C. This image illustrates the very large-scale coarsening that has been reached after 1 h 30 min of thermal annealing in the absence of compatibilization. (From T. S. Omonov, *Crucial Aspects of Phase Morphology Generation and Stabilization in Two- and Three-Phase Polymer Blends: Physical, Reactive and Combined Routes of Compatibilization*, Ph.D. thesis, Katholieke Universiteit Leuven, Belgium, 2007, under the supervision of C. Harrats and G. Groeninckx. For a detailed scientific discussion refer to T. S. Omonov, C. Harrats, G. Groeninckx, and P. Moldenaers, *Polymer* 48, 5289, 2007.)

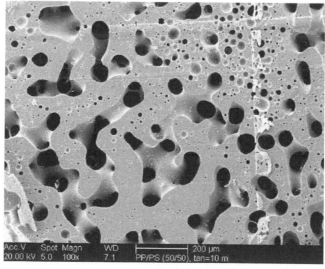

(A)

FIGURE 5.33

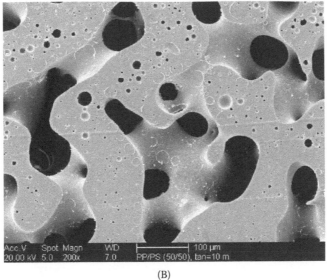

(B)

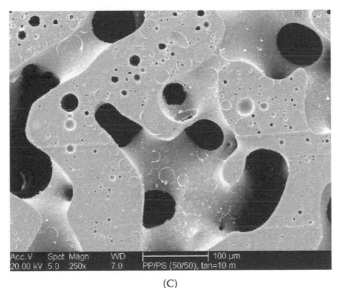

(C)

FIGURE 5.33 (*Continued*)
Different illustrations of SEM photomicrographs of cryosmoothed and chloroform-etched surface of melt-blended uncompatibilized 50 wt% polypropylene/50 wt% polystyrene blend thermally annealed for 10 min at 200°C. (A): 100× magnification. (B): 200× magnification. (C): 250× magnification. (From T. S. Omonov, *Crucial Aspects of Phase Morphology Generation and Stabilization in Two- and Three-Phase Polymer Blends: Physical, Reactive and Combined Routes of Compatibilization*, Ph.D. thesis, Katholieke Universiteit Leuven, Belgium, 2007, under the supervision of C. Harrats and G. Groeninckx.)

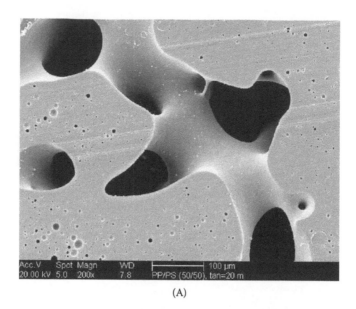

(A)

(B)

FIGURE 5.34

Different illustrations of SEM photomicrographs of cryosmoothed and chloroform-etched surface of melt-blended uncompatibilized 50 wt% polypropylene/50 wt% polystyrene blend thermally annealed for 20 min at 200°C. (A): 200× magnification. (B): 96× magnification. (C): 50× magnification. (Note: do not pay attention to the visible foreign fibers on the surface of observation.) (From T. S. Omonov, *Crucial Aspects of Phase Morphology Generation and Stabilization in Two- and Three-Phase Polymer Blends: Physical, Reactive and Combined Routes of Compatibilization*, Ph.D. thesis, Katholieke Universiteit Leuven, Belgium, 2007, under the supervision of C. Harrats and G. Groeninckx.)

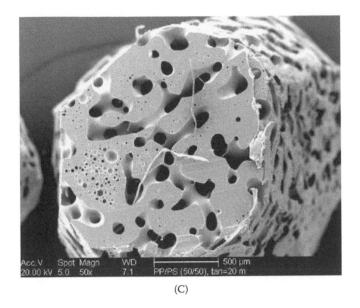

(C)

FIGURE 5.34
(*Continued*)

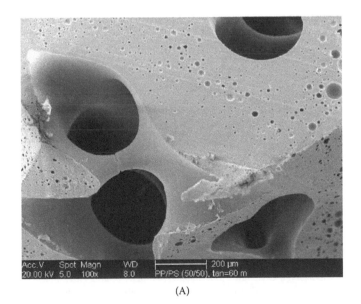

(A)

FIGURE 5.35

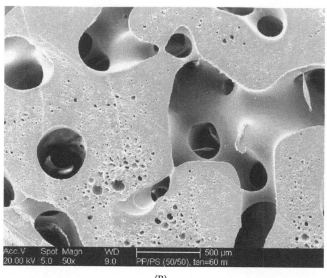

(B)

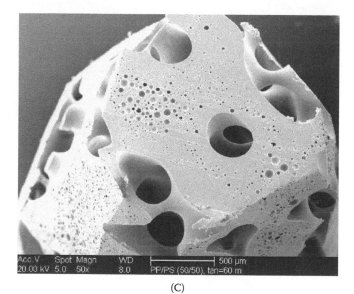

(C)

FIGURE 5.35 (*Continued*)
Series of illustrations of SEM photomicrographs of cryosmoothed and chloroform-etched surface of melt-blended uncompatibilized 50 wt% polypropylene/50 wt% polystyrene blend thermally annealed for 60 min at 200°C. (A): 100× magnification. (B): 50× magnification. (C): another view at 50× magnification. (T. S. Omonov, *Crucial Aspects of Phase Morphology Generation and Stabilization in Two- and Three-Phase Polymer Blends: Physical, Reactive and Combined Routes of Compatibilization*, Ph.D. thesis, Katholieke Universiteit Leuven, Belgium, 2007, under the supervision of C. Harrats and G. Groeninckx.)

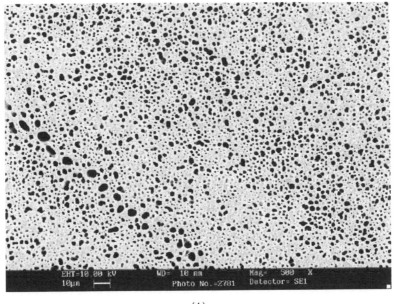

(A)

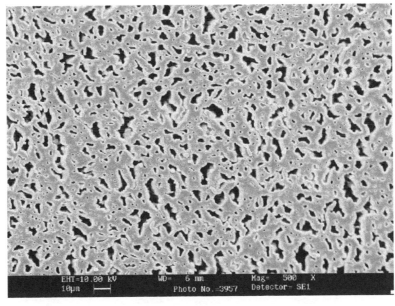

(B)

FIGURE 5.36

Effect of different components in multicomponent blends based on polyamide 6 (PA 6) and styrenic polymers. Addition of a reactive compatibilizer maleic anhydride grafted styrene acrylonitrile (SANMA) changed the matrix-dispersed particle structure of PA6/styrene acrylonitrile (SAN) to a cocontinuous one. Replacement of part of SAN by polybutadiene particles led to a coarse cocontinuous structure (PA6/acrylobutadiene styrene [ABS]), which was refined after compatibilizer addition. SEM of cut surfaces etched in different solvents. (A): PA6/SAN = 50/50 wt%, SAN etched in tetrahydrofurane; (B): PA6/ABS = 50/50 wt%, PA6 etched in formic acid; (C): PA6/ABS = 50/50 wt%, SAN etched in tetrahydrofurane; (D): PA6/SAN + SANMA, PA6 etched in formic acid; (E): PA6/ABS + SANMA, PA6 etched in formic acid; and (F): PA6/ABS = 50/50 wt%, PA6 etched in formic acid. (From S. H. Jafari, P. Pötschke, M. Stephan, G. Pompe, H. Warth, and H. Alberts, *Polymer* 43, 6985–6922, 2002. With permission.)

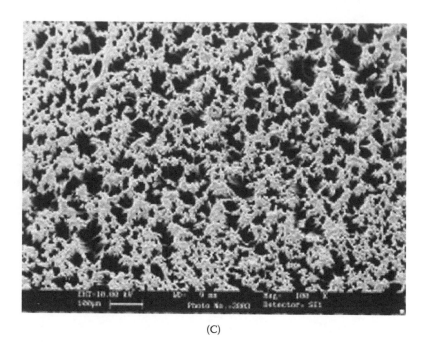

(C)

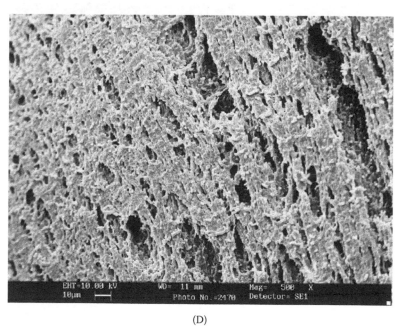

(D)

FIGURE 5.36
(*Continued*)

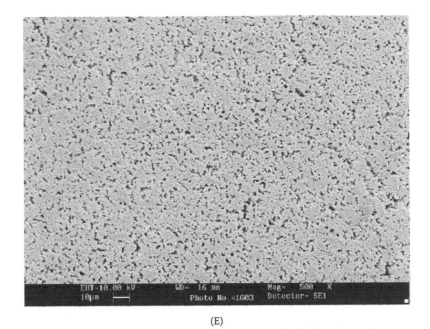

(E)

(F)

FIGURE 5.36
(*Continued*)

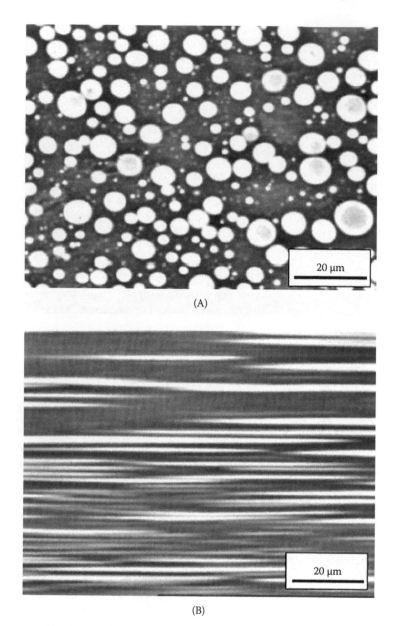

(A)

(B)

FIGURE 5.37

Morphology changes during melt elongation and subsequent recovery in polystyrene (PS)/poly(methyl acrylate) (PMMA) blends. Blends were extruded using a Brabender twin-screw extruder and compression molded. Morphology was investigated using light microscopy in phase contrast (PS darker phase) on thin section before and after stretching and during recovery. The images illustrate the recovery of the initial morphology type. (A, B, C, D): PS/PMMA = 65/35 vol.% (A: initial morphology; B: elongation 30 sec at a straining rate of 3; C: recovery time: 300 sec; D: recovery time 7320 sec). (E, F, G, H): PS/PMMA = 50/50 vol.% (E: initial morphology; F: elongation 30 sec at a straining rate of 3; G: recovery time: 300 sec; H: recovery time 7320 sec). (From U. Handge and P. Pötschke, *J. Rheol.* 84, 1103–1122, 2004. With permission.)

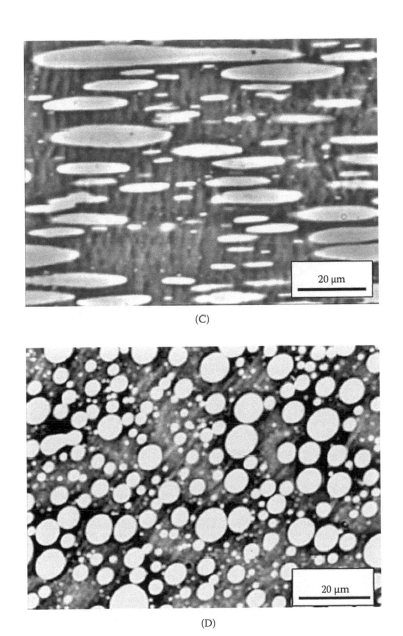

(C)

(D)

FIGURE 5.37
(*Continued*)

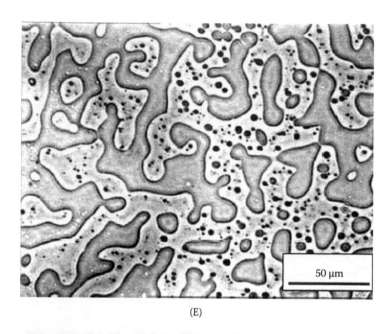

(E)

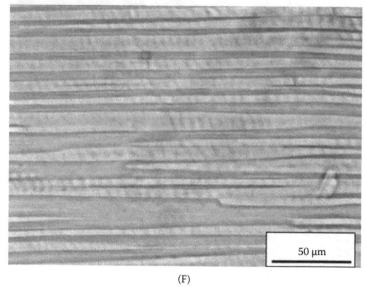

(F)

FIGURE 5.37
(*Continued*)

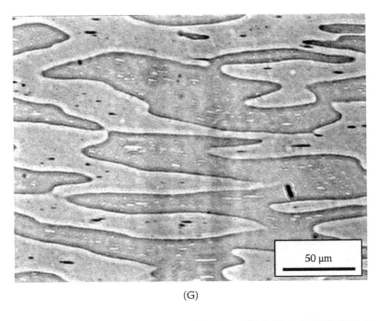

(G)

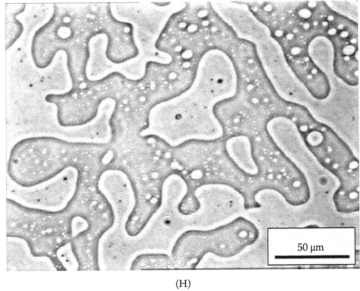

(H)

FIGURE 5.37
(*Continued*)

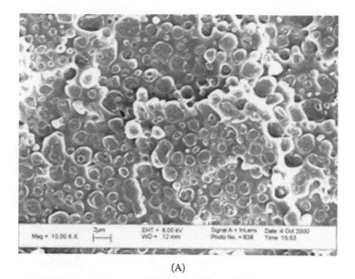

(A)

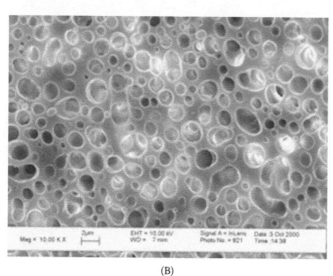

(B)

FIGURE 5.38
Illustrations of difficulties in unambiguously detecting cocontinuous structures in multiphase blends: Example PA6/SAN = 70/30 blends, extruded using Killion single-screw extruder at 80 rpm. The photomicrographs in (A: cryofracture), (B: cryosmoothed surface and THF etched), and (C: transmission electron microscopy [TEM] PA6 stained, taken perpendicular to the extruded strand direction) suggest a particle-in-matrix structure; whereas those in (D and E: SEM of cryofracture surfaces) clearly indicate a stretched cocontinuous structure (sample fractured along extrusion direction). (From U. Handge and P. Pötschke, *J. Rheol.* 84(5), 1103–1122, 2004. With permission.)

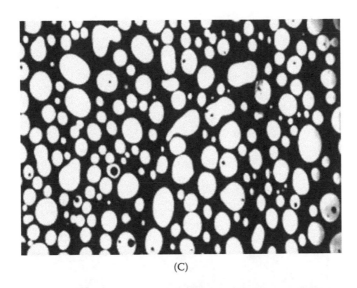

(C)

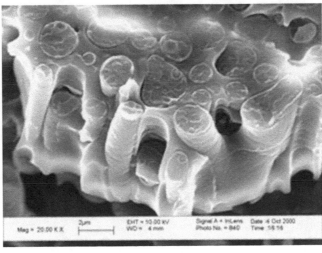

(D)

FIGURE 5.38
(*Continued*)

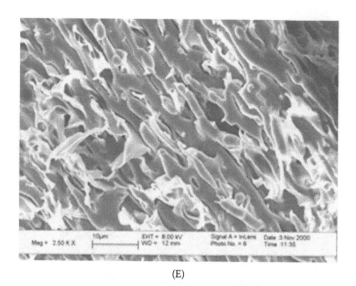

(E)

FIGURE 5.38
(*Continued*)

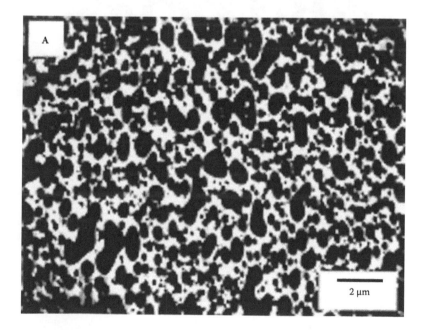

FIGURE 5.39
Morphology of extruded (PPE/PS)/SAN blends as observed by transmission electron microscopy. (A): PPE/SAN 60/40. (B): (PPE/PS)/SAN (45/15)/40. (C): (PPE/PS)/SAN (36/24)/40. (D): (PPE/PS)/SAN (30/30)/40. (E): (PPE/PS)/SAN (24/36)/40. (F): (PPE/PS)/SAN (15/45)/40. The photomicrographs were transferred to black-and-white images to allow computer-aided evaluation of the blend morphology. The black phase indicates the PPE/PS phase, whereas the white phase denotes the SAN phase. The microstructure of melt-processed, immiscible poly(2,6-dimethyl-1,4-phenylene ether)/poly(styrene-co-acrylonitrile) blends (PPE/SAN) was controlled by systematically adjusting the viscosity ratio between both phases. For this purpose, low-viscous polystyrene (PS) was added as a third component, as it shows selective miscibility with PPE and thus allows varying of the shear viscosity of PPE over a broad range. (From A. Göldel, H. Ruckdäschel, A. H. E. Müller, P. Pötschke, and V. Altstädt, *e-Polymer*, Nov. 2007. With permission.)

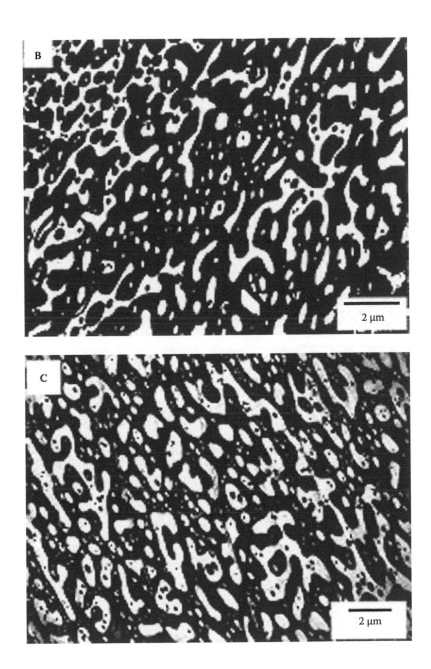

FIGURE 5.39
(*Continued*)

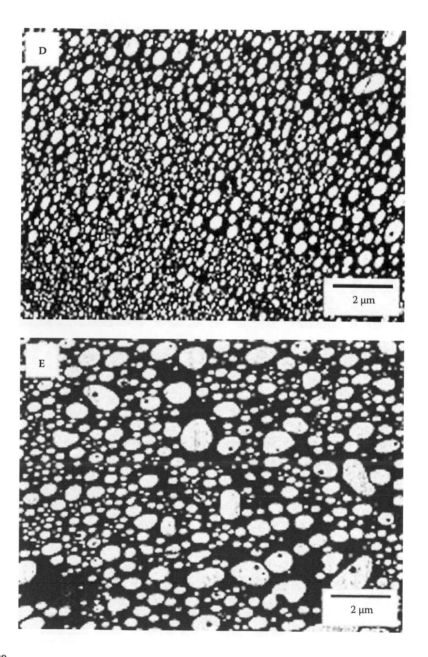

FIGURE 5.39
(*Continued*)

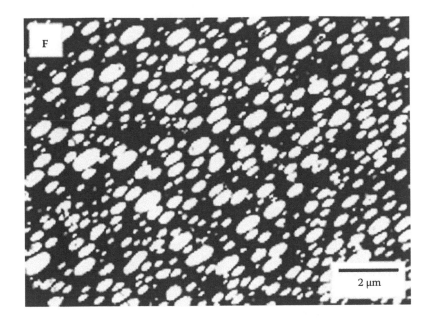

FIGURE 5.39
(*Continued*)

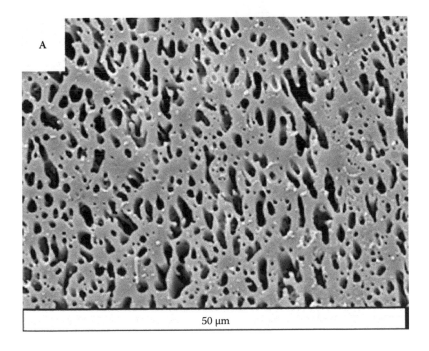

FIGURE 5.40
SEM photomicrographs of cryosmoothed and chloroform-etched surfaces of melt-blended PP/PS (50/50) blends. (A): without annealing; (B): after 5 min; (C): after 10 min; (D): after 20 min; and (E): after 60 min of annealing at 205°C. (From T. S. Omonov, *Crucial Aspects of Phase Morphology Generation and Stabilization in Two- and Three-Phase Polymer Blends: Physical, Reactive and Combined Routes of Compatibilization*, Ph.D. thesis, Katholieke Universiteit Leuven, Belgium, 2007, under the supervision of C. Harrats and G. Groeninckx.)

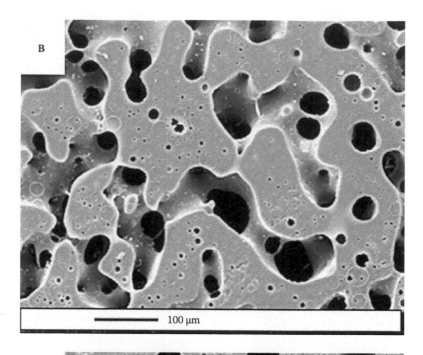

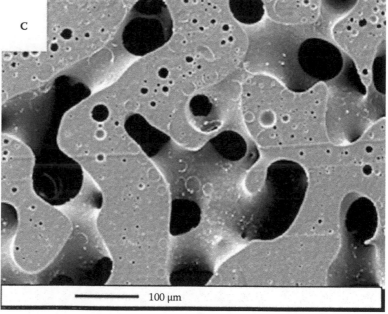

FIGURE 5.40
(*Continued*)

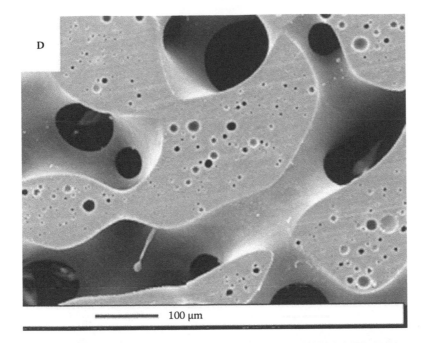

100 μm

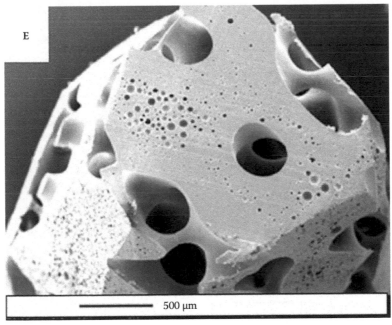

500 μm

FIGURE 5.40
(*Continued*)

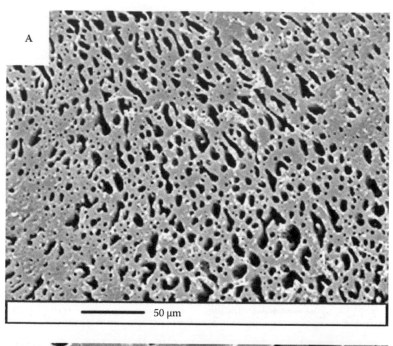

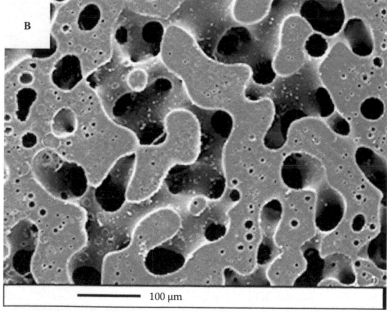

FIGURE 5.41
SEM photomicrographs of cryosmoothed and chloroform-etched surfaces of PP/PS (50/50) blends with 1 wt% (PP-g-MA1/PS-NH2). (A): without annealing; (B): after 5 min; (C): after 10 min; (D): after 20 min; and (E): after 60 min of annealing at 205°C. (From T. S. Omonov, *Crucial Aspects of Phase Morphology Generation and Stabilization in Two- and Three-Phase Polymer Blends: Physical, Reactive and Combined Routes of Compatibilization*, Ph.D. thesis, Katholieke Universiteit Leuven, Belgium, 2007, under the supervision of C. Harrats and G. Groeninckx.)

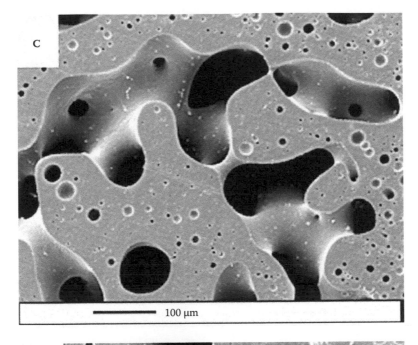

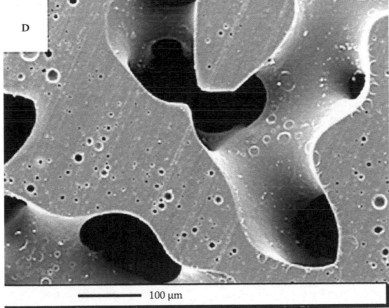

FIGURE 5.41
(*Continued*)

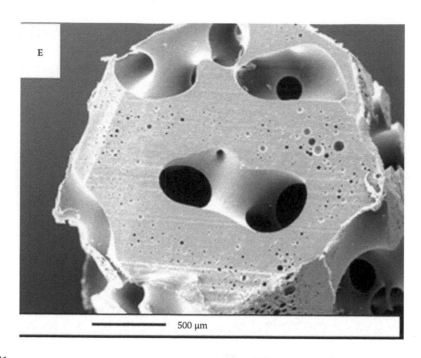

FIGURE 5.41
(*Continued*)

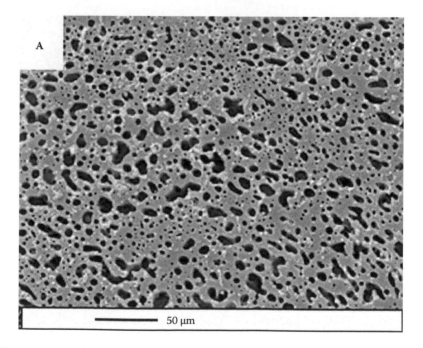

FIGURE 5.42
SEM of cryosmoothed and chloroform-etched surfaces of melt-blended PP/PS (50/50) blends with 5 wt% (PP-g-MA1/PS-NH2). (A): without annealing; (B): after 5 min; (C): after 10 min; (D): after 20 min; and (E): after 60 min of annealing at 205°C. (From T. S. Omonov, *Crucial Aspects of Phase Morphology Generation and Stabilization in Two- and Three-Phase Polymer Blends: Physical, Reactive and Combined Routes of Compatibilization*, Ph.D. thesis, Katholieke Universiteit Leuven, Belgium, 2007, under the supervision of C. Harrats and G. Groeninckx.)

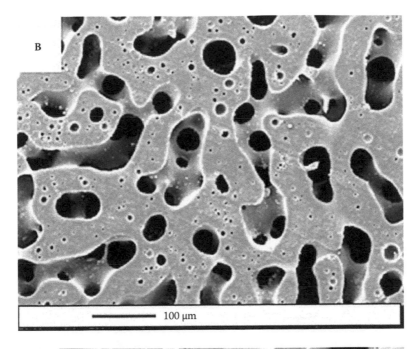

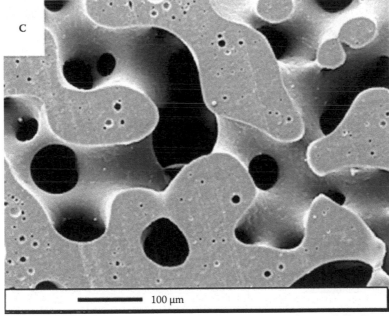

FIGURE 5.42
(*Continued*)

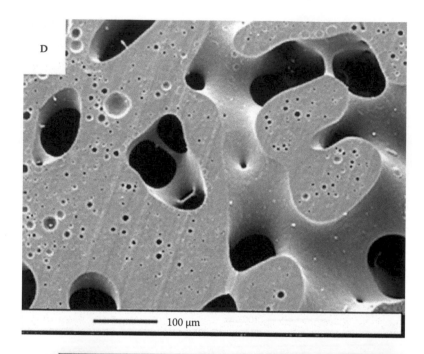

D

100 μm

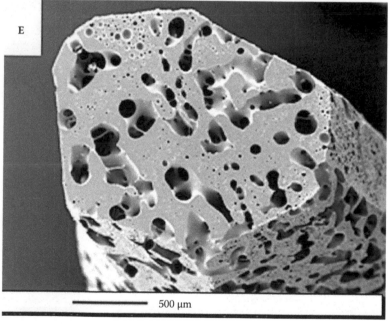

E

500 μm

FIGURE 5.42
(*Continued*)

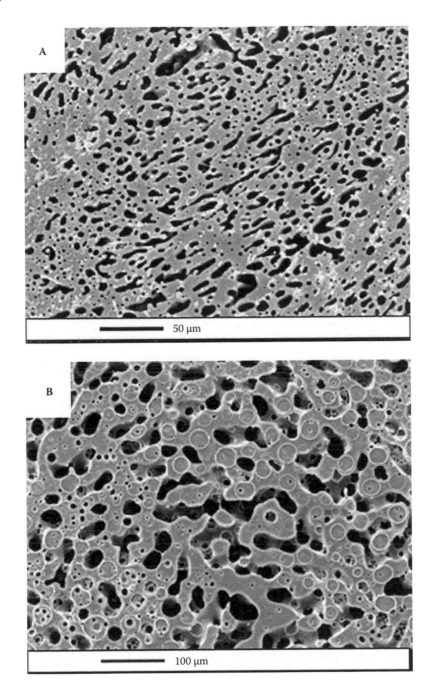

FIGURE 5.43

SEM of cryosmoothed and chloroform-etched surfaces of melt-blended PP/PS (50/50) blends with 1 wt% (PP-g-MA7.8/PS-NH2). (A): without annealing; (B): after 5 min; (C): after 10 min; (D): after 20 min; and (E): after 60 min of annealing at 205°C. (From T. S. Omonov, *Crucial Aspects of Phase Morphology Generation and Stabilization in Two- and Three-Phase Polymer Blends: Physical, Reactive and Combined Routes of Compatibilization*, Ph.D. thesis, Katholieke Universiteit Leuven, Belgium, 2007, under the supervision of C. Harrats and G. Groeninckx.)

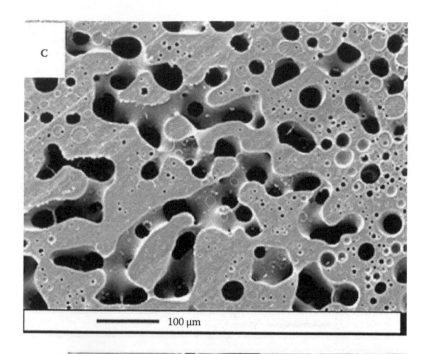

100 μm

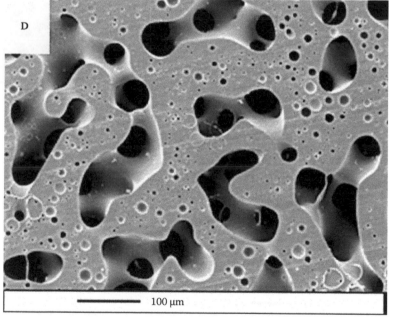

100 μm

FIGURE 5.43
(*Continued*)

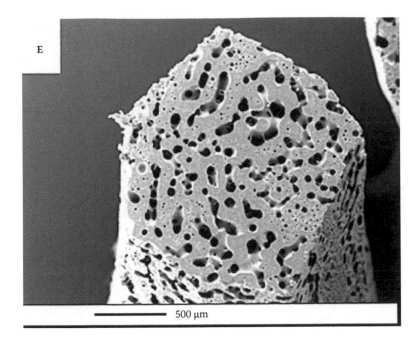

FIGURE 5.43
(*Continued*)

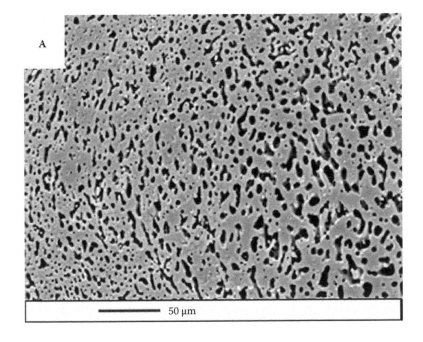

FIGURE 5.44
SEM photomicrographs of cryosmoothed and chloroform-etched surfaces of melt-blended PP/PS (50/50) blends with 5 wt% (PP-g-MA7.8/PS-NH2). (A): without annealing; (B): after 5 min; (C): after 10 min; (D): after 20 min; and (E): after 60 min of annealing at 205°C. (From T. S. Omonov, *Crucial Aspects of Phase Morphology Generation and Stabilization in Two- and Three-Phase Polymer Blends: Physical, Reactive and Combined Routes of Compatibilization*, Ph.D. thesis, Katholieke Universiteit Leuven, Belgium, 2007, under the supervision of C. Harrats and G. Groeninckx.)

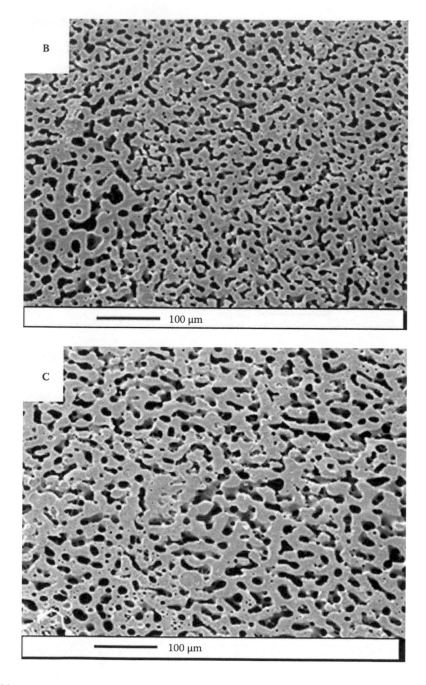

FIGURE 5.44
(*Continued*)

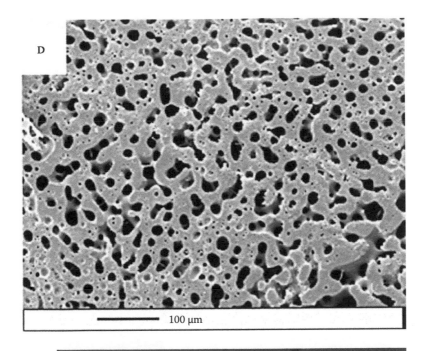

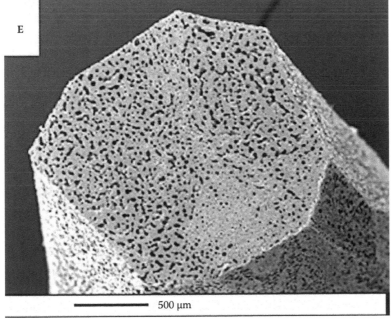

FIGURE 5.44
(*Continued*)

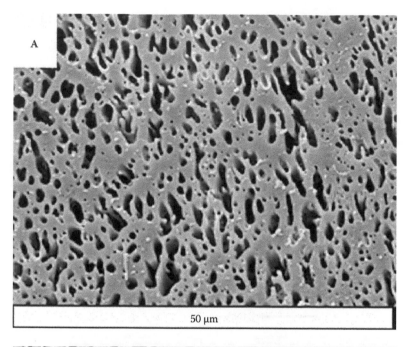

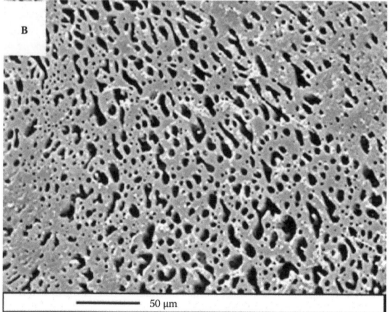

FIGURE 5.45
SEM photomicrographs (500× magnification) of cryosmoothed and chloroform-etched surfaces of melt-blended 50/50 PS/PP blend. (A): without copolymer; (B): + 1 wt% (PP-MA1 + PS-NH2); (C): + 5 wt% (PP-MA1 + PS-NH2); (D): + 1 wt% (PP-MA8 + PS-NH2); and (E): + 5 wt% (PP-MA8 + PS-NH2). (From T. S. Omonov, *Crucial Aspects of Phase Morphology Generation and Stabilization in Two- and Three-Phase Polymer Blends: Physical, Reactive and Combined Routes of Compatibilization*, Ph.D. thesis, Katholieke Universiteit Leuven, Belgium, 2007, under the supervision of C. Harrats and G. Groeninckx.)

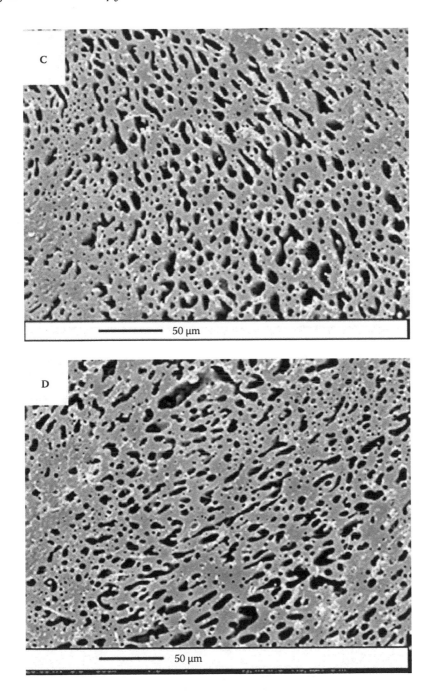

FIGURE 5.45
(*Continued*)

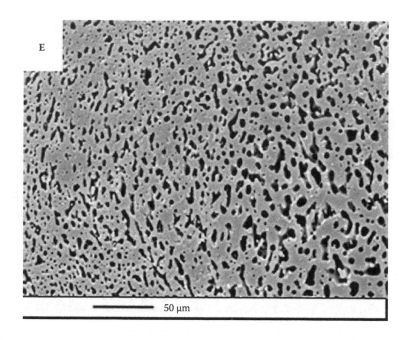

FIGURE 5.45
(*Continued*)

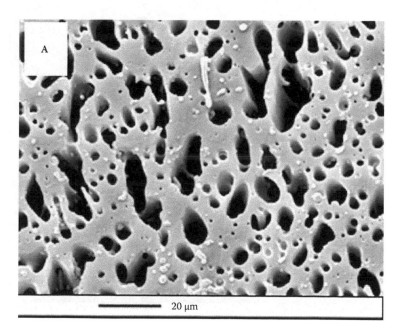

FIGURE 5.46
SEM photomicrographs (1000× magnification) of cryosmoothed and chloroform-etched surfaces of melt-blended 50/50 PS/PP blends: comparison of the effectiveness of the two compatibilizers at various concentrations to stabilize the phase morphology against phase coarsening. (A): without copolymer; (B): + 1 wt% (PP-MA1 + PS-NH2); (C): + 5 wt% (PP-MA1 + PS-NH2); (D): + 1 wt% (PP-MA8 + PS-NH2); and (E): + 5 wt% (PP-MA8 + PS-NH2). (From T. S. Omonov, *Crucial Aspects of Phase Morphology Generation and Stabilization in Two- and Three-Phase Polymer Blends: Physical, Reactive and Combined Routes of Compatibilization*, Ph.D. thesis, Katholieke Universiteit Leuven, Belgium, 2007, under the supervision of C. Harrats and G. Groeninckx.)

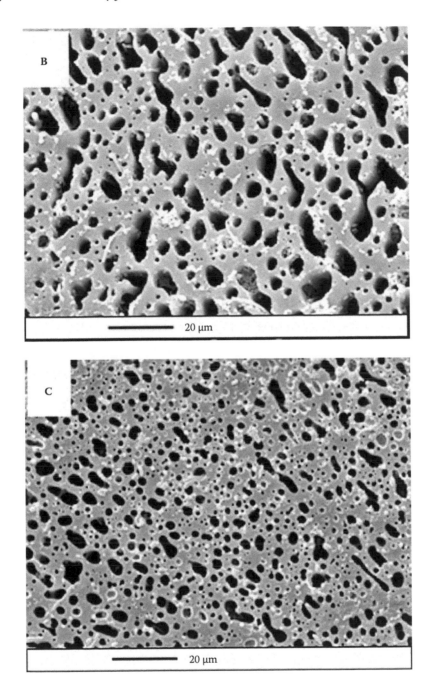

FIGURE 5.46
(*Continued*)

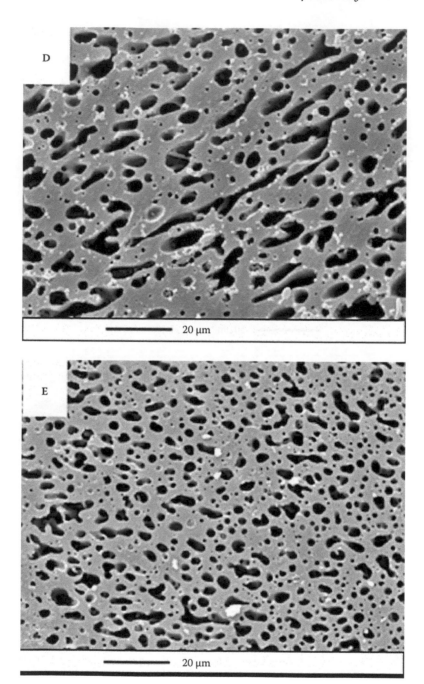

FIGURE 5.46
(*Continued*)

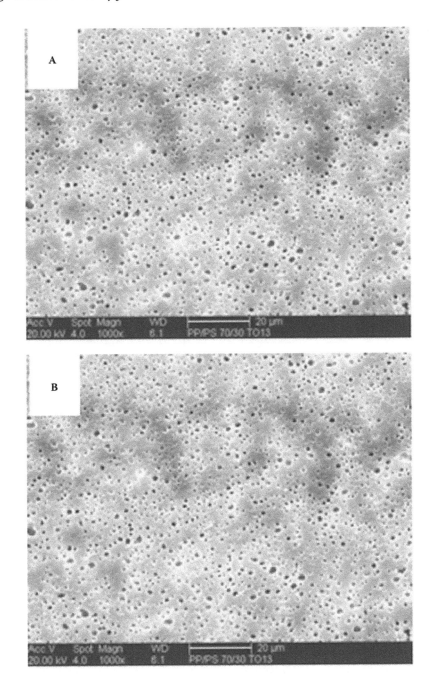

FIGURE 5.47
SEM photomicrographs of cryosmoothed and chloroform-etched surfaces of melt-blended PP/PS blends in perpendicular direction to extrusion flow. (A): 80/20; (B): 70/30; (C): 60/40; (D): 50/50; (E): 40/60; (F): 30/70; and (G): 20/80. Note that the blends rich in PP exhibit an apparent droplet-in-matrix phase morphology when observed in the transverse to extrusion direction, whereas it is fully continuous in the opposite compositions. As detailed in article cited below, the apparent dispersed phase morphology is cocontinuous but exhibits a highly elongated (oriented) structure. (From T. S. Omonov, *Crucial Aspects of Phase Morphology Generation and Stabilization in Two- and Three-Phase Polymer Blends: Physical, Reactive and Combined Routes of Compatibilization*, Ph.D. thesis, Katholieke Universiteit Leuven, Belgium, 2007, under the supervision of C. Harrats and G. Groeninckx.)

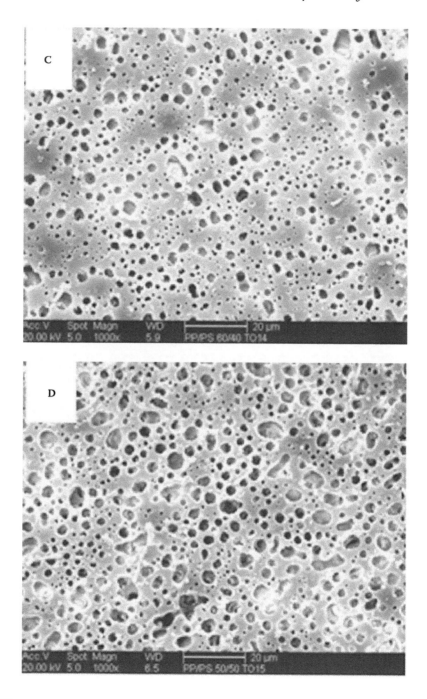

FIGURE 5.47
(*Continued*)

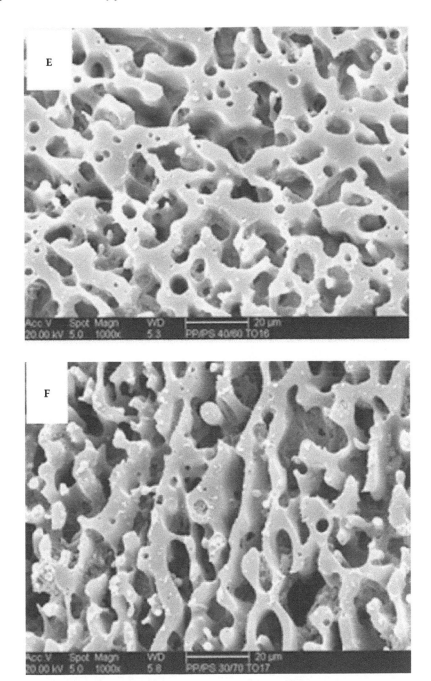

FIGURE 5.47
(*Continued*)

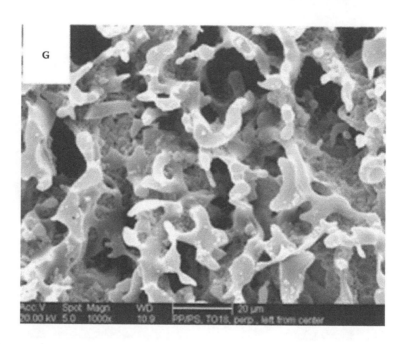

FIGURE 5.47
(*Continued*)

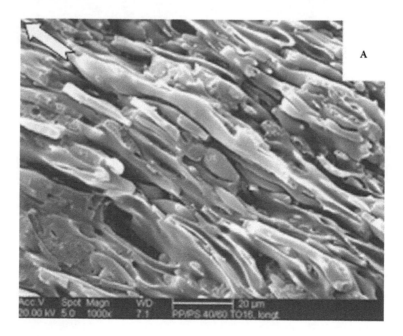

FIGURE 5.48
SEM photomicrographs of cryofractured surfaces of melt-blended PP/PS blends in a parallel direction to extrusion flow. (A): 80/20; (B): 70/30; (C): 60/40; (D): 50/50; (E): 40/60; (F): 30/70; and (G): 20/80. Note that, in contrast to Figure 5.47, the phase morphology in this direction reveals the elongated structure of the minor phase. (From T. S. Omonov, *Crucial Aspects of Phase Morphology Generation and Stabilization in Two- and Three-Phase Polymer Blends: Physical, Reactive and Combined Routes of Compatibilization*, Ph.D. thesis, Katholieke Universiteit Leuven, Belgium, 2007, under the supervision of C. Harrats and G. Groeninckx.)

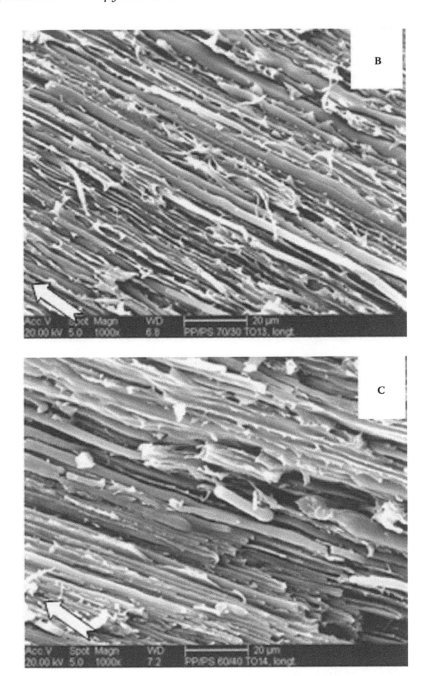

FIGURE 5.48
(*Continued*)

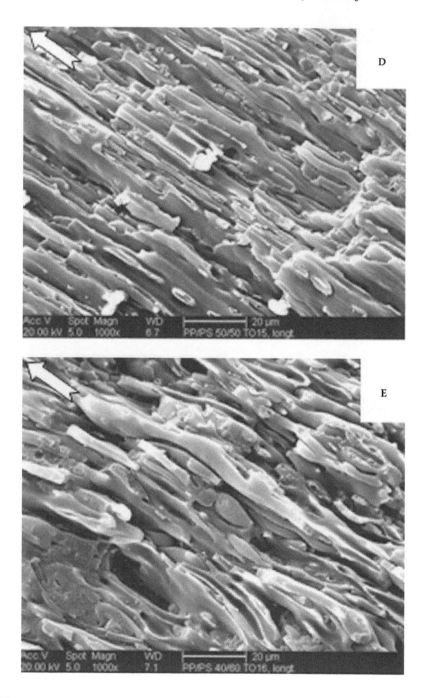

FIGURE 5.48
(*Continued*)

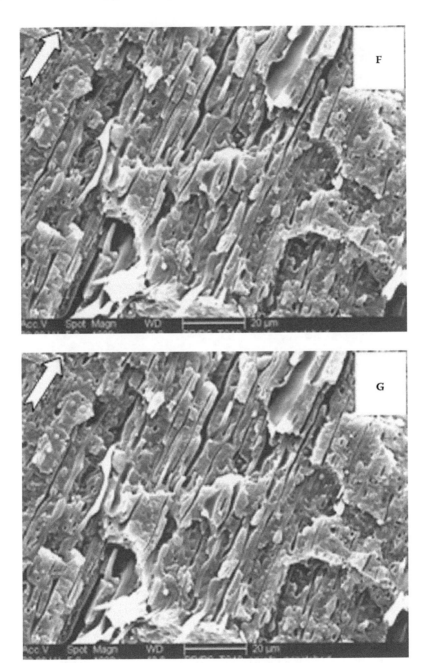

FIGURE 5.48
(*Continued*)

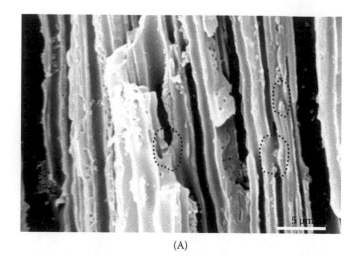

(A)

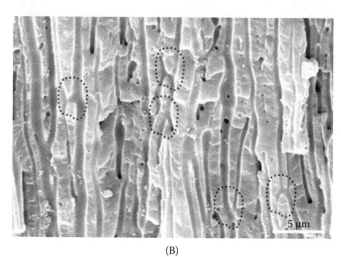

(B)

FIGURE 5.49

SEM photomicrographs of cryofractured and chloroform-etched surfaces cut parallel to the extrusion direction. (A): blend containing 70 wt% PP; and (B): blend containing 80 wt% PP. All the PCHMA phase contained in the 70/30 and 80/20 PP/PCHMA blends was selectively extracted, leaving the PP phase as a unique, integral, and self-supporting piece in the solvent. After morphological observations were performed on the surfaces of these blends cut in the parallel direction with respect to the extrusion flow, a really oriented, elongated, and infinite rod-like structure was found, as seen in the two images. The question that arises from this morphology is how the continuity of these elongated rod-like structures of the PCHMA minor phase is ensured. A close observation of the images presented reveals that these long rods are interconnected in two possible modes—the most likely is via multiple connections as indicated by the dashed lines on the SEM micrographs, and also via small (0.2 mm diameter) and short connecting rods acting as bridges between the long and continuous major rods (see the small holes on the images). (From C. Harrats, T. S. Omonov, G. Groeninckx, and P. Moldenaers, *Polymer* 45, 8115–8126, 2004. With permission.)

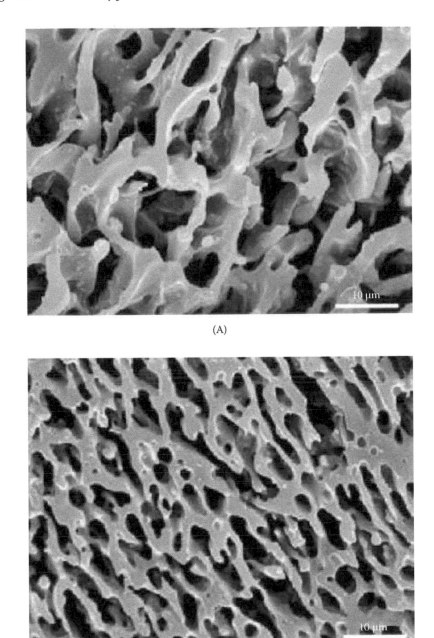

(A)

(B)

FIGURE 5.50

SEM photomicrographs of cryosmoothed and chloroform-etched surfaces of melt-blended PP/PCHMA blends containing the following: (A): 30; (B): 40; (C): 50; (D): 70; and (E): 80 wt% PP. (From C. Harrats, T. S. Omonov, G. Groeninckx, and P. Moldenaers, *Polymer* 45, 8115–8126, 2004. With permission.)

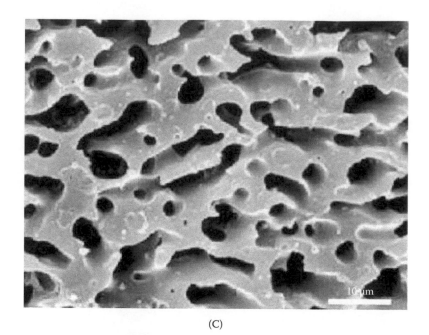

(C)

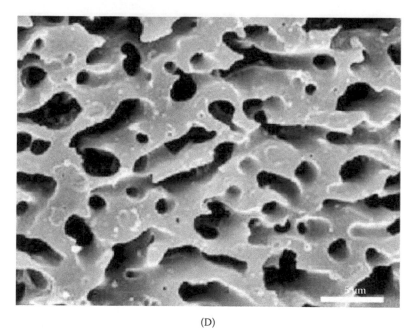

(D)

FIGURE 5.50
(*Continued*)

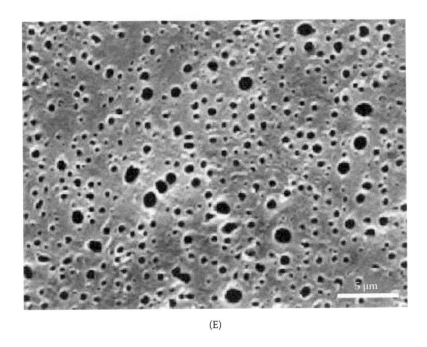

(E)

FIGURE 5.50
(*Continued*)

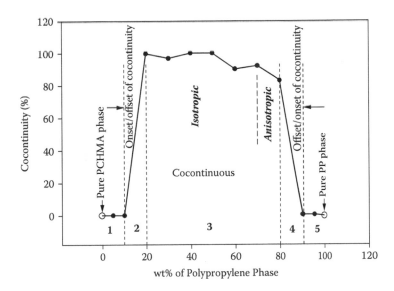

FIGURE 5.51
Percentage of cocontinuity as a function of PP content in the uncompatibilized PP/PCHMA blends showing the various composition windows where the phase morphology is dispersed, partially cocontinuous, or fully cocontinuous. (From C. Harrats, T. S. Omonov, G. Groeninckx, and P. Moldenaers, *Polymer* 45, 8115–8126, 2004. With permission.)

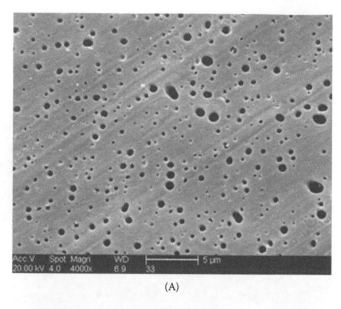

(A)

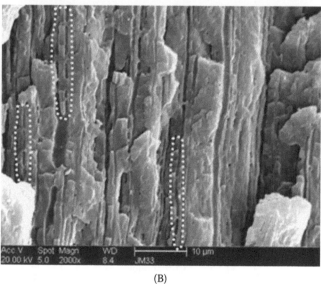

(B)

FIGURE 5.52

SEM photomicrographs of 80 PP/20 PCHMA +1 wt% copolymer based on PP-MA8 precursor. (A): cryosmoothed and chloroform-etched surfaces cut perpendicular to the extrusion direction; and (B): cryofractured and chloroform-etched surfaces cut parallel to the extrusion direction. (From C. Harrats, T. S. Omonov, G. Groeninckx, and P. Moldenaers, *Polymer* 45, 8115–8126, 2004. With permission.)

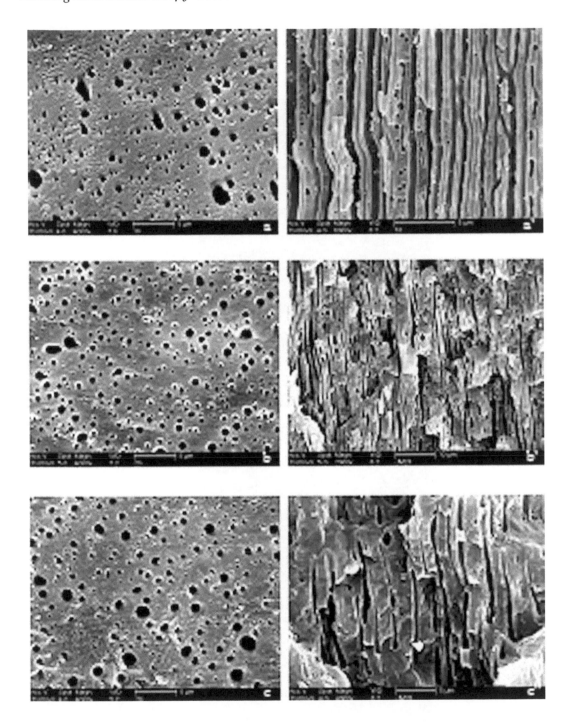

FIGURE 5.53
SEM photomicrographs of 80 PP/20 PCHMA + (a, a': 1; b, b': 5; and c, c': 10 wt%) copolymer based on PP-MA1 precursor. Right series: cryosmoothed and chloroform-etched surfaces cut perpendicular to the extrusion direction; left series: cryofractured and chloroform-etched surfaces cut parallel to the extrusion direction. (From C. Harrats, T. S. Omonov, G. Groeninckx, and P. Moldenaers, *Polymer* 45, 8115–8126, 2004. With permission.)

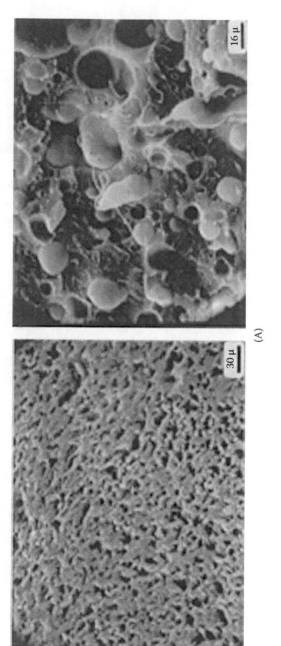

(A)

FIGURE 5.54

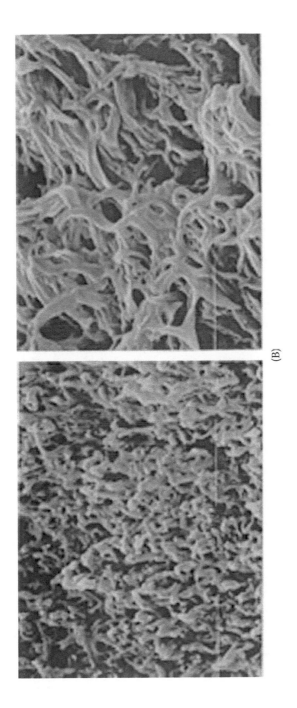

(B)

FIGURE 5.54
(*Continued*)

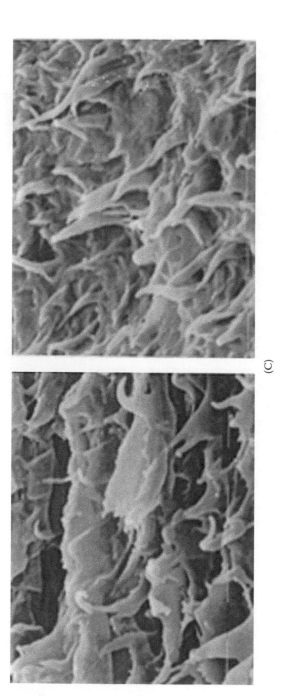

(C)

FIGURE 5.54 (*Continued*)
SEM of fractured and THF-etched surfaces of 20 low-density polyethylene (LDPE)/80 PS. Comparison of efficiency of copolymers in stabilizing the cocontinuous phase morphologies upon annealing at 180°C for 150 min. (A): without copolymer; (left) nonannealed, (right) annealed. (B): + 10 wt% (Tp) pure diblock; (left) nonannealed, (right) annealed. (C): + 10 wt% (Pr) tapered diblock; (left) nonannealed, (right) annealed. (From C. Harrats, R. Fayt, R. Jérôme, and S. Blacher, *J. Polym. Sci.: Part B: Polym. Phys.* 41, 202–216, 2003. With permission.)

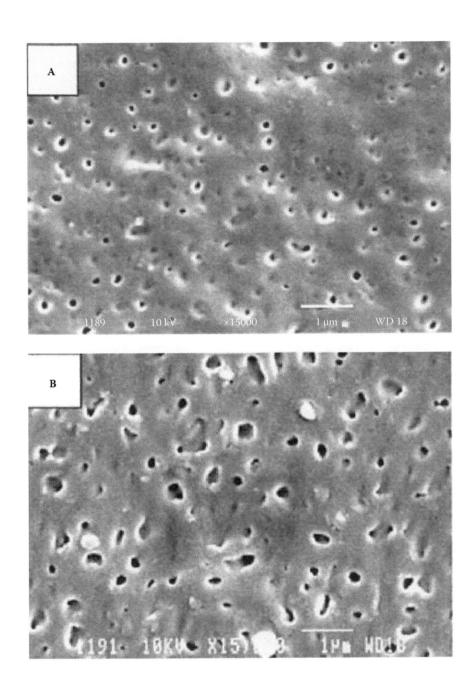

FIGURE 5.55

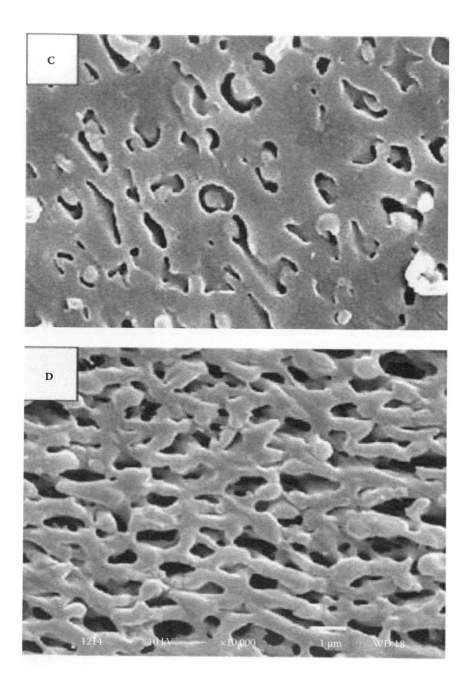

FIGURE 5.55
(*Continued*)

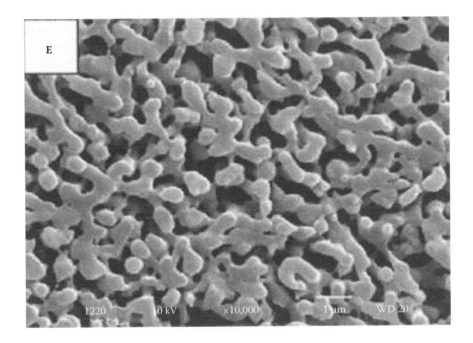

FIGURE 5.55 (*Continued*)
SEM photomicrographs of melt-blended PMMA/polycarbonate PC blends extracted with formic acid. Extracted microtomed samples were prepared for qualitative analysis of the blend microstructure. First, microtomy was performed with a Leica 2165 microtome equipped with a glass knife. Then, the PMMA was selectively extracted by soxhlet extraction with formic acid for 48 h. Finally, dried samples were gold–palladium coated and SEM micrographs were obtained with a JEOL 840 microscope at 10 kV. The white bar indicates 1 μm. (A): 20% of PMMA, 15,000×; (B): 30% of PMMA, 15,000×; (C): 40% of PMMA, 8500×; (D): 50% of PMMA, 10,000×; and (E): 60% of PMMA, 10,000×. The SEM micrographs indicate a dispersed phase in matrix morphology for compositions from 10 to 40% of PMMA and a cocontinuous morphology for 50 and 60%. (From N. Marin and B. D. Favis, *Polymer* 43, 4723–4731, 2002. With permission.)

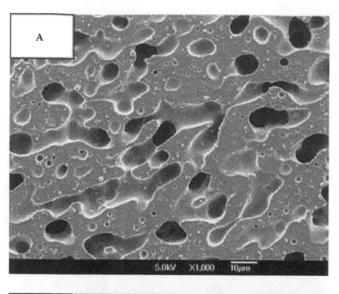

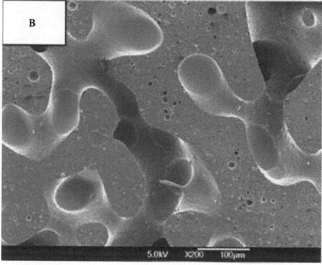

FIGURE 5.56

SEM micrographs of 50/50 polystyrene/polyethylene polymer blends with various amounts of 40k PS–PE copolymer, prepared by mixing at 50 rpm for 10 min at 170°C. The micrographs (A, C, E, G) represent blends quenched immediately after mixing, whereas the micrographs (B, D, F, H) represent blends annealed for 15 min at 170°C. (A and B) have no block copolymer; (C and D) have 0.1%; (E and F) have 0.3%; and (G and H) have 1% (short scale bar = 10 μm; longer one = 100 μm). The phase size of the quenched samples decreases as the PS–PE concentration increases from 0 to 1%. Annealing the blends with 0.1% block copolymer for 15 min results in coarsening of the morphology, but the phase size is much smaller than the neat blend as shown in (F and D). Increasing the concentration to 0.3% yields better suppression of coarsening. However, the morphology after 15 min of annealing is coarser than the blend with 1% of 40k PS–PE copolymer (F and H). (From J. A. Galloway, H. K. Jeon, J. R. Bell, and C. W. Macosko, *Polymer* 46, 183–191, 2005. With permission.)

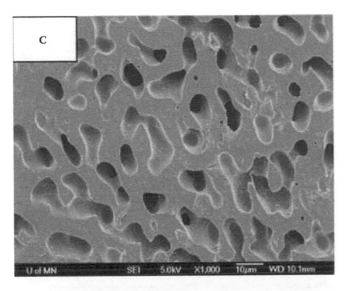

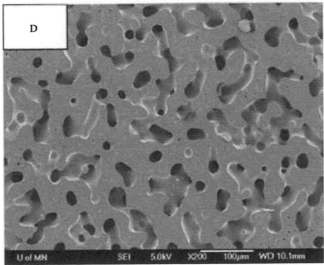

FIGURE 5.56
(*Continued*)

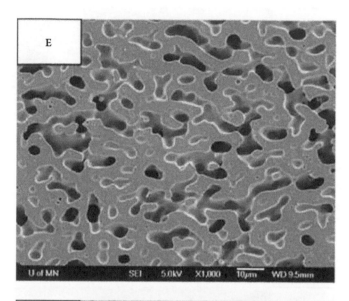

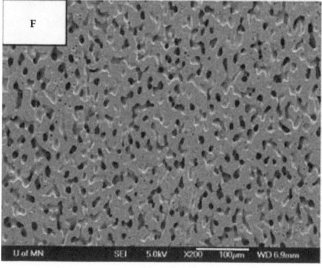

FIGURE 5.56
(*Continued*)

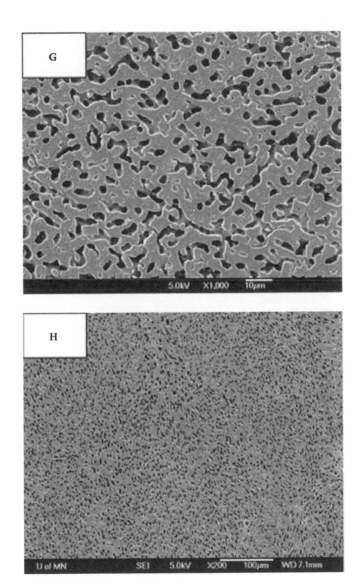

FIGURE 5.56
(*Continued*)

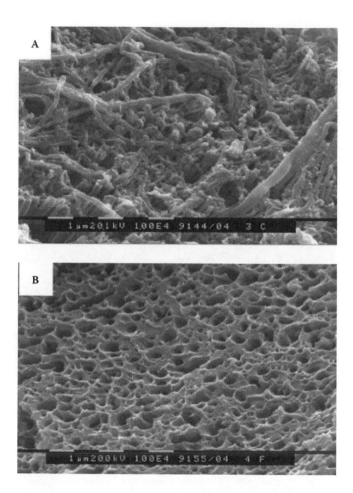

FIGURE 5.57

SEM images of the PMMA phase after PS extraction (A, C, E, G) and the PS phase after PMMA extraction (B, D, F, H) in 10 min melt-mixed PS/PMMA blends. (A and B): PMMA 30%; (C and D): PMMA 40%; (E and F): PMMA 60%; and (G and H): PMMA 75%. A bicontinuous phase structure is observed in all four blends in the range 30 to 75% of PMMA irrespective of whether the PS or PMMA phase has been extracted, indicating phase cocontinuity in this volume fraction interval. Furthermore, the micrographs of PMMA and PS at the same PMMA content match quite well. The PMMA phase of the 20% PMMA sample and the PS phase of the 85% PMMA sample were spongy and collapsed during extraction but remained as one coherent structure (not presented here). However, the PMMA phase of the 10% PMMA sample (extracted PS phase) and the PS phase of the 90% PMMA samples (extracted PMMA phase) collapsed to a powder during extraction and no SEM images were obtained. It is interesting to observe in (A and H) that the minority phases in the cocontinuous blends are stretched out into fibril-like structures. (From C. Z. Chuai, K. Almdal, and J. Lyngaae-Jørgensen, *Polymer* 44, 481–493, 2003. With permission.)

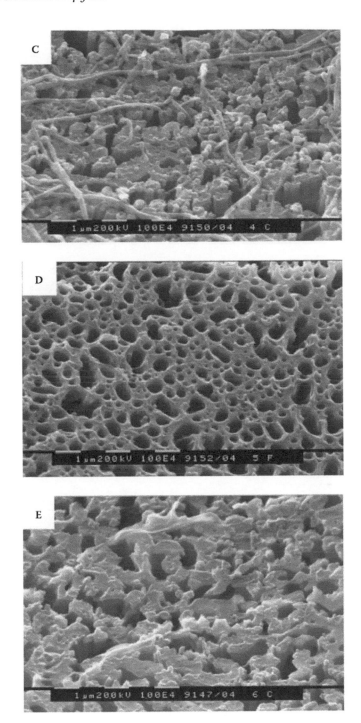

FIGURE 5.57
(*Continued*)

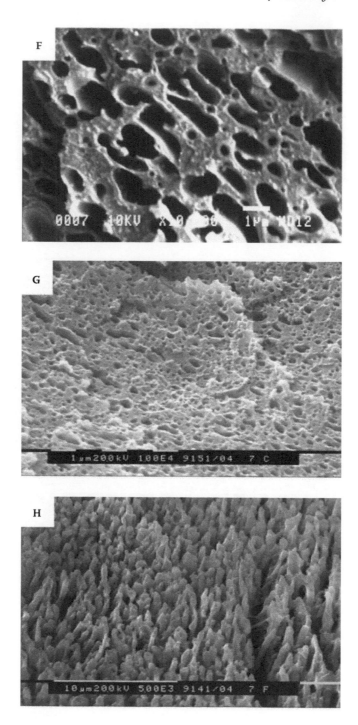

FIGURE 5.57
(*Continued*)

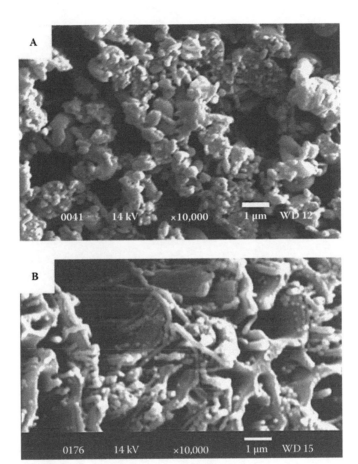

FIGURE 5.58
SEM images of the PMMA phase after PS extraction and the PS phase after PMMA extraction in PS/PMMA blends for 10 min mixing. (A and B): 40% PMMA; (C and D): 60% PMMA; (E and F): 70% PMMA; and (G and H): 80% PMMA. Note that the polystyrene grade used here has a lower viscosity than the PS used in Figure 5.57. A cocontinuous structure is observed in all four blends in the range 40 to 80% PMMA irrespective of whether the PS or PMMA phase has been extracted, indicating phase cocontinuity in this composition interval. Furthermore, the micrographs of PMMA and PS at the same PMMA content match quite well. It is interesting to note that the 40% PMMA shows a coalescent morphology of PMMA droplets (A). The PMMA phases of the 30% PMMA sample and the PS 85% PMMA sample were spongy and collapsed during extraction but remained as one coherent structure (not shown). The PMMA phase of the 20% PMMA sample and the PS phase of the 90% PMMA sample disintegrated to powders during extraction, indicating a droplet-in-matrix phase morphology. (From C. Z. Chuai, K. Almdal, and J. Lyngaae-Jørgensen, *Polymer* 44, 481–493, 2003. With permission.)

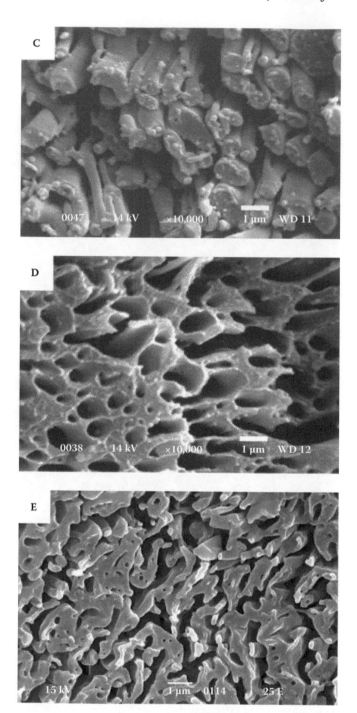

FIGURE 5.58
(*Continued*)

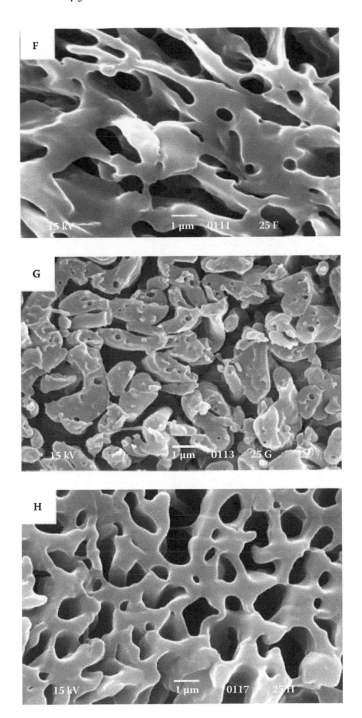

FIGURE 5.58
(*Continued*)

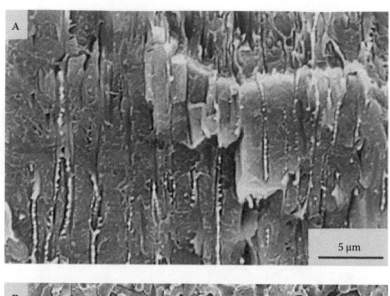

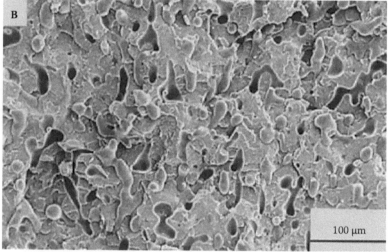

FIGURE 5.59

SEM micrographs for annealed polystyrene/polyethylene 70/30 blend. (A): 0 min; (B): 1 min; (C): 16 min; and (D): 75 min. The micrograph in (A) has a magnification of 5000×, whereas the other micrographs (B through D) have a magnification of 200×. The micrograph in (A) (0 min) shows the morphology of the blend obtained after its molding. Polystyrene forms a partially continuous structure with polyethylene. After 1 min of annealing and subsequent quenching in liquid nitrogen, the morphology obtained is drastically disrupted and an irregular-shaped dispersed phase (PE) is observed. This can be attributed to the partial retraction and breakup of the polyethylene phase, which becomes the dispersed phase in this case. The time required for the breakup of polyethylene layers can be roughly estimated given the viscosity of both components (PS and PE) and their interfacial tension via Tomotika's equation for Newtonian fluids. For 16 min annealing and longer, a PE droplet–PS matrix morphology is observed, with coalescence probably occurring at longer annealing times. It is likely that the cocontinuous structure at this composition is formed during the elongational flow in the die rather than in the mixing zone (kneading blocks) of the extruder. The nonequilibrium morphology obtained after cooling in the water trough can be highly affected upon annealing the blend due to viscoelastic effects as well as interfacial tension. (From N. Mekhilef, B. D. Favis, and P. J. Carreau, *J. Polym. Sci. Part B: Polym. Phys.* 35, 293–308, 1997. With permission.)

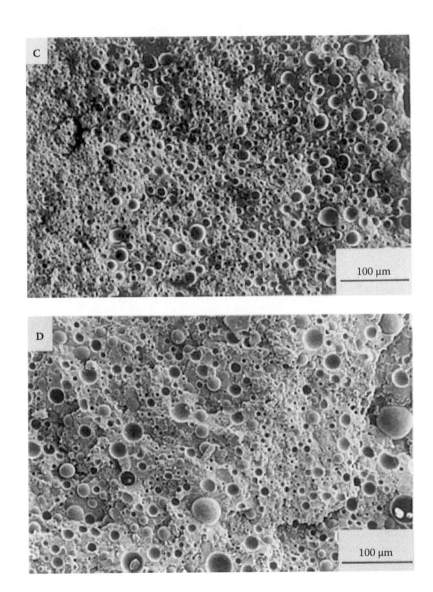

FIGURE 5.59
(*Continued*)

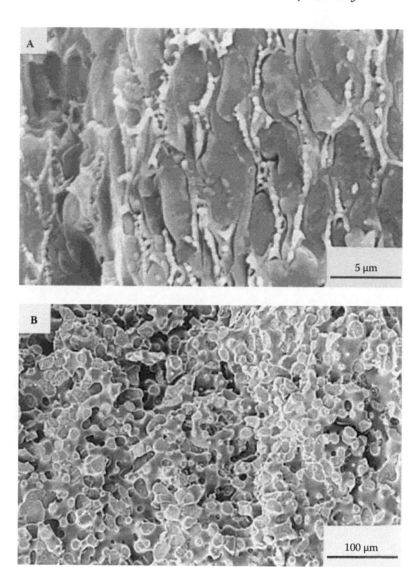

FIGURE 5.60
SEM micrographs for the annealed PS/PE 50/50 blend. (A): 0 min; (B): 1 min; (C): 16 min; and (D): 75 min. The morphology of the annealed 50/50/PS/PE blend is shown for the same magnification as for Figure 5.59A through Figure 5.59D. In this case, the initial morphology observed in (A) shows a continuous structure for both polystyrene and polyethylene in the core of the sample. As the blend is annealed, a significant coarsening is observed with increasing annealing time. However, cocontinuity is maintained and both phases form a continuum. According to the magnification of the micrographs, the change in the fiber diameter of this blend is on the order of 25 times. It is interesting to note that at this specific composition (50/50), the original cocontinuous morphology is maintained but coarsens with annealing time. This suggests that true dual-phase continuity exists at this composition or at compositions close to 50/50. (From N. Mekhilef, B. D. Favis, and P. J. Carreau, *J. Poly. Sci. Part B: Polym. Phys.* 35, 293–308, 1997. With permission.)

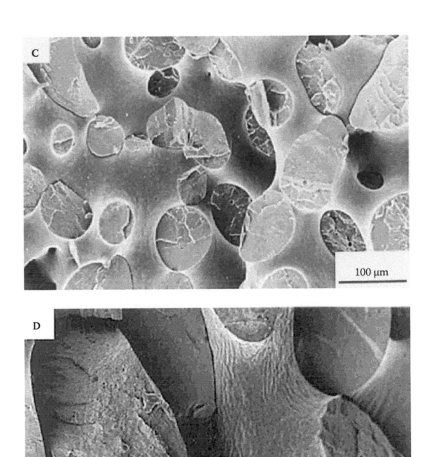

FIGURE 5.60
(*Continued*)

Index

A

Acrylobutadiene styrene (ABS), 152, 153, 154, 263
Anisotropic phase morphology, 26, 222
Atomic force microscopy (AFM), 30

B

Bentonite clay, 50, 52
Blend composition, 3–4
Breakup, 2, 3, 13, 160, 168, 176, 221, 228, 322
   in a blend composition, 3–4
   and coalescence, 4
Butadiene styrene, 43, 193, 263
Butanediol dimethacrylate (BDDM), 135, 136
Butyl methacrylate (BMA), 135, 136

C

Carbon nanotubes (CNT), 115
Clay
   aspect ratio, 35
   exfoliation, 34
   silicate, 37, 41
Cloisite organoclay, 34, 36, 46, 47, 48, 49
Coalescence, 1, 3, 9, 10
   in binary polymer blends, 4–5
   and breakup, 4
   and mean field approach, 12
Cocontinuity, by PP content, 303
Cocontinuous morphologies
   in melt-mixed polymer blends, 13
   in multiphase blends, 270–272
   in partially miscible blends, 10–12
   thermoplastic/thermoplastic blends, 12
   thermoplastic/thermoset blends, 12–13
   thick, 8
   thin, 9
Compatibilization, 5, 9
Composite droplet phase, 7, 10–13, 217, 218
Compound microscope, 17

D

Deformation, 1, 2, 20
   avoiding, 165
   due to lack of cooling, 19
   and elasticity, 6
   and shear stress, 6, 195
DGEBA, see Di-glycidyl ether of bisphenol A
   (DGEBA)
Diacrylate diisocyanate (DAD), 157

Dicyandiamide (DICY), 169
Di-glycidyl ether of bisphenol A (DGEBA), 169, 170,
   171, 172, 173
   DGEBA/MDEA, 158, 160, 165, 168
Dispersed morphologies, 7, 9–10, 219
DMA (n-dodecyl methacrylate), 135, 136
Droplet-in-matrix phase, see Dispersed morphologies

E

Elasticity, 5–6
Electron microscope, 18
Elongational flow fields, 1, 2, 3, 7, 9, 10, 322
Epoxy-acrylobutadienestyrene (ABS) blends,
   152–154
Epoxy-amine, 162
Epoxy thermosets, 169
Etching
   before and after, 126
   plasma-etched surface, 28, 30, 100, 102
   selective, 25, 185, 241, 251
   with tetrahydrofuran (THF), 131, 148, 150, 192, 270,
   308
   with xylene, 28
Ethylene-propylene-rubber (EPR), 210, 211
Ethylene vinyl acetate (EVA), 34, 35, 36, 37, 87, 88, 89,
   90, 91, 92
Ethyl methacrylate (EMA), 135, 136
EVA, see Ethylene vinyl acetate (EVA)
Extruders
   single-screw, 6
   twin-screw, 6–7

H

Harkin's equation, 10
HDPE, see High-density polyethylene (HDPE)
High-density polyethylene (HDPE), 210, 211, 217,
   218, 219
High-impact polystyrene (HIPS), 10
Hooke, 17
2 hydroxy ethyl acrylate (HEA), 156, 157
Hyperbolic (elongational) flow fields, 1, 2, 3, 7, 9, 10,
   322
Hyperbranched polymer (HBP), 131, 132, 151

I

Injection molding, 6
Internal mixers, 6
Interpenetrating network (IPN), 135, 136

J

Janssen, Zacharias, 17

L

Layered silicates, 33, 36, 37
LDPE, *see* Low-density polyethylene (LDPE)
Leeuwenhoek, Antoni van, 17
Light microscope, 18
Low-density polyethylene (LDPE), 94, 95, 126, 127,
        128, 210, 211
    LDPE/PMMA blends, 174–175
    LDPE/PS blend, 175–176, 192, 308–310

M

Maleic anhydride grafted styrene acrylonitrile
        (SANMA), 263–265
Maleic anhydride grafted styrene-ethylene
        butylene-styrene (SEBS-g-MA)/SEBS, 215–216
Maleic anhydride modified polypropylene (MAPP),
        61, 62, 63
Mean field approach, 12
Methacrylic acid, 94, 169
Methylenebis(2,6-diethylaniline) (MDEA), 158
Methylenedianiline (MDA), 169
Methyl methacrylate (MMA), 46, 47, 51, 135,
        136, 170, 171
Methyltetrahydrophthalic anhydride (MTHPA), 169
Micrographia (Hooke), 17
Microscopic tools, 17, 18
Mixing equipment, 6–7
Mixing parameters, 6
Montmorillonite (MMT), 33, 36, 37, 38, 39, 40, 41, 67,
        69, 71, 73, 75, 77, 79, 80, 82, 94, 95, 99; *see also*
        Polypropylene (PP)
MWNT (multiwalled carbon nanotubes), 97, 98, 108,
        109, 110, 111, 112, 113

N

Nanocomposites, 51, 84
    clay aspect ratio, 35
    intercalation/exfoliation, 33
    from LDPE, 95, 96
    limitations, 33–34
    from organoclay, 93, 94
    polymer-clay interface, 35
    preparation techniques, 34
    properties, 33, 35
Nanofillers, 36
Nanotubes, 97, 98, 110, 111, 112, 115
N-butyl methacrylate (BMA), 135
N-dodecyl methacrylate (DMA), 135, 136, 170, 171
Newtonian fluids, 7
N-isopropyl acrylamide (NIPAM), 53

Nylon 6, 43–45, 85, 213
Nylon/clay, 33

O

Organophilized montmorillonite clay
        (OMMT), 61
Oxazoline-terminated PP (PP-Ox), 131, 151
Oxazoline-terminated PS (PS-Ox), 131, 151

P

PA6, *see* Polyamide 6 (PA6)
PCL, *see* Poly-ε-caprolactone (PCL)
PEO, *see* Polyethylene oxide (PEO)
Peroxide, 135, 136
PET, *see* Polyethylene terephthalate (PET)
Plasma-etched surface, 28, 30, 100, 102
PMMA, *see* Polymethyl methacrylate (PMMA)
PMMA-NH2, 133, 134, 137
Polyamide 6 (PA6), 2, 13, 37, 85, 86, 116, 117, 118,
        119, 120, 176, 178, 180, 181, 182, 183, 185, 186,
        188, 190, 263
    PA6/acrylobutadiene styrene (ABS), 263–265
    PA6/polypropylene (PP), 178–179, 181
    PA6/PP/polystyrene (PS), 183, 184–185, 186–187
    PA6/PP/ PP-MA8, 182
    PA6/PP/PP–MA2/PS/SMA2, 188–189
    PA6/(PP/PP–MA8)/(PS/SMA2) blends, 190–191
    PA6/PS, 176–177
    PA6/PS/SMA2 blends, 180
    PA6/SAN, 270–272
    PA6/styrene acrylonitrile (SAN), 263–265
Polybutadien, 5, 10, 169, 174, 195, 263
Polycaprolactone (PCL), 22, 54
    PCL/PEO blends, 56, 57, 58, 59, 60
Polycarbonate, 6, 12, 97, 98, 108, 110, 111, 112, 113,
        115, 311
Polycyclohexylmethacrylate, 9, 20, 21, 185
Polydimethylsiloxane, 195
Polydimethylsiloxane/polybutadiene, 195–197
Poly(divinyl benzene), 53
Poly-ε-caprolactone (PCL), 58, 59, 60
Polyethylene/copolyamide (PE/CPA), 124, 125
Polyethylene oxide (PEO), 12, 22, 34, 37, 54, 55, 57,
        58, 59, 60
Polyethylene terephthalate (PET), 123, 125, 126,
        127, 128
Polylactide (PLA), 198–200, 201, 203, 204
Polymer-clay interface, 35
Polymer nanocomposites, 33; *see also* Sample
        preparation
Polymethyl methacrylate (PMMA), 5, 219, 316–318,
        319–321
    PMMA-NH2, 133, 134, 137
    PMMA (Plexiglas 7H) nanocomposite, 48, 49
    PMMA/polycarbonate (PC) blends, 311

Polyoelfins, 33, 211–212
Polyphenylene ether (PPE), 171, 172, 173, 272
    PPE/PS/SAN blends, 272–275
Polyphenylene oxide (PPO), 10, 116, 117, 118
    PPO/PA6, 119, 120
Poly(propene-g-styrene) graft copolymers, 148–150
Polypropylene oxide, 156, 157
Polypropylene oxide/2,4 toluene diisocyanate/
        2- hydroxyethyl acrylate (TDI/HEA)
        precursor, 156
Polypropylene (PP), 20, 84
    PP/MMT/polypyrrole (PPy), 100, 101, 102, 103,
        104, 105
    PP/montmorillonite (MMT)/PC, 114, 115
    PP/OMMT/maleic anhydride modified
        polypropylene (MAPP), 61, 62
    PP/organophilized montmorillonite clay
        (OMMT), 61
    PP/PCHMA (80/20), 304, 305–307
    PP/PCHMA blends, 300, 301–303, 303
    PP/polystyrene (PS) 2/1 blends, 146, 147, 148–150,
        150–151
    PP/PS (20/80) blends, 230–232, 233–235, 242, 243,
        244, 245, 296, 299
    PP/PS (30/70) blends, 226, 227, 228, 229, 295, 299
    PP/PS (40/60) blends, 225, 236–237, 295, 298
    PP/PS (50/50) blends, 224, 225, 238, 239, 241,
        246–247, 248–249, 252, 259, 260–261, 262,
        275–277, 278–279, 280–282, 283–284, 285–287,
        294, 298
    PP/PS (60/40) blends, 223, 294, 297
    PP/PS (70/30) blends, 221, 222, 293, 297
    PP/PS (80/20) blends, 293, 296
    PP/PS/styrene-butadiene-styrene (SBS), 193–194
Polypyrrole (PPy), 102, 105
Polystyrene (PS), 64, 240, 254, 266, 272; *see also*
        Polypropylene (PP)
    high-Mw polystyrene/high-Mw PMMA, 218
    low-Mw polystyrene/low-Mw PMMA, 217
    PS/DGEBA/MDEA, 158–159, 160–161, 163–165,
        166–168
    PS/epoxy-amine 60/40 blend, 162
    PS/PM, 137, 140, 141
    PS/PM/SCA, 138, 139, 142, 143, 144, 145
    PS/polyethylene (PE) 50/50 blend, 312–315,
        324–325
    PS/polyethylene (PE) 70/30 blend, 322–323
    PS/polymethyl acrylate (PMMA) blends, 266–269
    PS/PP (50/50), 250, 251, 253, 255, 256, 257, 258,
        288–290, 290–292
    PS/SBR-2/PE blends, 207–208
    PS/SBR/PO, 209–210, 211–212
    PS/styrene-butadiene rubber 1 (SBR-1)/PE blends,
        206–207
Polystyrene (PS)/M₃(C₁₈)₁ composite, 67, 68, 79
Polyvinyl alcohol, 37
Polyvinylidene fluoride (PVDF), 37, 38, 39, 40, 41
PPE, *see* Polyphenylene ether (PPE)

PP-g-Ox/PS-COOH (70/30) blends, 154–155
PP-MA1 precursor, 305
PP-MA8 precursor, 304
PPO, *see* Polyphenylene oxide (PPO)
PPO 2000/isophorone diisocyanate (IPDI)/HEA, 157
PP-Ox, 131, 132
PP-Ox/PS-Ox, 151–152
PPy, *see* Polypyrrole (PPy)
PS-COOH, 133, 134, 137, 150, 154
PS-COOH/PMMA-NH2 (PS/PM), 133, 134, 137
PS-Ox, 131, 132
PS/PM, *see* PS-COOH/PMMA-NH2 (PS/PM)

S

Sample preparation
    cyrosurfaces, 19–20, 25
    and electrical conduction, 18
    lack of cooling, 19, 20
    metal coating, 18, 19
    mounting, 18
    of nonconducting materials, 18
    particle-in-matrix adhesion, 20–21, 22, 23, 24
SAN, *see* Styrene acrylonitrile (SAN)
Scanning electron microscopy (SEM), 18
SEBS, *see* Styrene-ethylene-butylene-styrene (SEBS)
Shear-flow fields, 2, 9
Shear stress, 6
    and deformation, 6, 195
Silane-containing coupling agent (SCA), 133, 134, 137
Silicone rubber, 43
Styrene acrylonitrile (SAN), 64–66, 80–81, 263, 270, 272
    SAN-2/M₃(C₁₈)₁ composite, 69, 70
    SAN-13.5/M₃(C₁₈)₁ composite, 71, 72, 81
    SAN-25/M₃(C₁₈)₁ composite, 73, 74, 83
    SAN-38/M₃(C₁₈)₁ composite, 75, 76
    SAN-58/M₃(C₁₈)₁ composite, 77, 78
Styrene-butadien-rubber (SBR), 206, 208, 210, 211
Styrene ethylene butylene-graft-diethyl maleate, 123
Styrene-ethylene-butylene-styrene (SEBS), 10
    SEBS-g-DEM, 123, 126, 128
    SEBS-g-MA, 213, 214, 215, 216
    SEBS-g-MA/a-PA blends, 213–214
Styrene maleic anhydride (SMA)
    SMA/M₃(C₁₈)₁ composite, 80, 82
Sulfonated syndiotactic polystyrene (SsPS), 106, 107
Surface artifacts, 23

T

Talc, 36, 37
Ternary blends, 7, 8, 10, 24, 25
Tetrahydrofuran (THF), 131, 148, 150, 192, 270, 308
Thermoplastic polyurethane (TPU), 27–30, 129, 130
Thermoplastics, 6
Thermoplastic starch (TPS), 198–200
    TPS/PLA blends, 201, 202, 203–204, 204–205

Thermoplastic/thermoplastic blends, 12
Thermoplastic/thermoset blends, 12–13
Thermosets, 169
Toluene diisocyanate (TDI), 156, 157
Transmission electron microscopy (TEM), 18
Trigonox-101, 135, 136

U

Ultra-fine full-vulcanized powdered rubber/
    montmorillonite (UFPRM), 43, 45
Ultra-fine full-vulcanized powdered rubber (UFPR),
    43–44, 45
Ultra-low-density polyethylene (ULDPE), 125, 127

V

Very low density polyethylene (VLDPE), 123
Vinyl pyridine, 43
Viscosity ratio, 1, 154, 272
    in binary blends, 3
    and critical capillary number, 2, 3
    dependence of phase size, 7
    and flow field, 2
    and melting temperature, 13

X

X-ray diffraction (XRD), 45